Technical Peer Review Report

Report of the Review Panel

Review of Selected Nuclear Safety Programs at Savannah River Site

ASME International

Institute for Regulatory Science

DISCLAIMER

This report was prepared through the collaborative efforts of the American Society of Mechanical Engineers (ASME) Center for Research and Technology Development and the Institute for Regulatory Science (referred to thereafter with the collaborators as the Society) for the Office of Science and Technology Development of the U.S. Department of Energy (referred to hereafter as the Sponsor).

Neither the Society nor the Sponsor, or others involved in the preparation or review of this report nor any of their respective employees, members, or persons acting on their behalf, make any warranty, expressed or implied, or assume any legal liability or responsibility for the accuracy, completeness, or usefulness of any information, apparatus, product, or process disclosed, or represent that its uses would not infringe privately owned rights.

Information contained in this work has been obtained by the American Society of Mechanical Engineers from sources believed to be reliable. However, neither ASME nor its authors or editors guarantee the accuracy or completeness of any information published in this work. Neither ASME nor its authors and editors shall be responsible for any errors, omissions, or damages arising out of the use of this information. The work is published with the understanding that ASME and its authors and editors are supplying information but are not attempting to render engineering or other professional services. If such engineering or professional services are required, the assistance of an appropriate professional should be sought.

Statement from By-Laws: The Society shall not be responsible for statements or opinions advanced in papers . . . or printed in its publications. (7.1.3)

ISBN No. 0-7918-0196-9

TABLE OF CONTENTS PAGE NO.

Preface

This report contains the results of a peer review performed jointly by the American Society of Mechanical Engineers (ASME) and the Institute for Regulatory Science (RSI). Based on a request from the U.S. Department of Energy (DOE), a Review Panel (RP) was established to peer review "Selected Nuclear Safety Programs at Savannah River Site". Keeping with the ASME/RSI process, the following RP was appointed by the Peer Review Committee for Energy and the Environment (PRCEE) of ASME:

James W. Dickey
William S. Durant
Jerome F. McMahon
Francis J. Patti
Richard Rosenberg

During the period covered by this report, the ASME PRCEE overseeing the peer review consisted of the following individuals:

Charles O. Velzy, Member of Executive Panel (EP) and Chair
Ernest L. Daman, Member of EP
Nathan H. Hurt, Member of EP
A. Alan Moghissi, Member of EP; Principal Investigator of the Peer Review Program
Gary A. Benda
Erich W. Bretthauer
Irwin Feller
Robert A. Fjeld
William T. Gregory, III
Peter B. Lederman
Jeffrey A. Marqusee
Lawrence C. Mohr, Jr.
Goetz K. Oertel
Glen W. Suter, II
Cheryl A. Trottier

The supporting staff included the following individuals:

Michael Tinkleman: Director of Research at the Center for Research and Technology Development of ASME in Washington, DC; and Administrative Manager of the ASME PRCEE.

Betty R. Love: Executive Vice President, RSI, Columbia, MD; and Administrative Manager of the Peer Review Program.

M.C. Kirkland: Vice President for Southeastern Region, RSI; and Project Manager.

Sorin R. Straja: Vice President for Science and Technology, RSI; and Technical Secretary.

Sharon D. Jones: Director of Training Programs, RSI; and Manager of Review Panel Operations.

The biographical summaries of the members of the RP, the PRCEE, and the technical staff are located at the end of this report.

The Review Criteria were provided by the Project Team. These criteria were slightly revised by the Technical Secretary and approved by the Project Team. The RP received appropriate documents describing various aspects of the project. The summary of the project included in this report was prepared by the Technical Secretary and approved by the Project Team. In addition, the staff of RSI undertook the task of preparing a listing of acronyms and abbreviations. This list, as reviewed and approved by the Project Team, is also included in this report.

The RP met from January 13, 2003, through January 17, 2003, in Augusta, GA. At the beginning of the meeting, the RP and the Project Team were introduced to the ASME/RSI peer review process and the DOE's desire for a nonconflicted and independent peer review. The introduction was followed by presentations by the Project Team and a discussion period. The agenda of this meeting is shown in the Appendix. At the end of the meeting, the RP met in an executive session to develop its strategy for writing the report. Subsequently, the RP wrote its findings and recommendations. The *Report of the Review Panel* was then copyedited.

Consistent with the procedures established by the ASME/RSI process, the copyedited *Report of the Review Panel* was provided to DOE for identification of potential errors, misunderstandings, and areas of ambiguity. Comments received by the DOE were considered by the RP and the report was revised accordingly.

Charles O. Velzy
A. Alan Moghissi

Appendix

ASME/RSI Peer Review
Review of Selected Nuclear Safety Programs at Savannah River Site

Augusta, GA – January 13-17, 2003

AGENDA

Monday, January 13, 2003

Radisson Riverfront Hotel Augusta

8:00 a.m.	ASME Peer Review Process	Michael Kirkland Institute for Regulatory Science
9:00 a.m.	Nuclear Safety Program (NSP) Overview NSP Roles and Responsibilities ***(Criteria 1 & 2)***	Andrew Vincent Westinghouse Savannah River Co.
10:00 a.m.	Break	
10:15 a.m.	NSP Roles and Responsibilities *(continued)* Nuclear Safety Training ***(Criterion 3)***	Andrew Vincent Chuck Voldness, WSRC Gibert Kennedy, WSRC Jim Junker Bechtel Savannah River, Inc.
12:00 noon	Lunch	
1:00 p.m.	Functional Classification ***(Criterion 4)*** USQ Process ***(Criteria 7 & 8)***	Bob Lowrie, WSMS Ed Hallinan, WSRC
3:00 p.m.	Break	
3:15 p.m.	Performance Metrics ***(Criteria 5 & 6)***	Chuck Voldness Ken Yates, WSRC
5:00 p.m.	Adjournment	

Tuesday, January 14, 2003

Radisson Riverfront Hotel Augusta

8:00 a.m.	Executive Session
10:00 a.m.	Discussion
12:00 noon	Lunch

ASME/RSI Peer Review
Review of Selected Nuclear Safety Programs at Savannah River Site

Augusta, GA – January 13-17, 2003

AGENDA

Tuesday continued

1:00 p.m. Discussion *(continued if necessary)*

3:00 p.m. Break

3:15 p.m. Executive Session

5:00 p.m. Adjournment

Wednesday, January 15, 2003

Radisson Riverfront Hotel Augusta

8:30 a.m. Writing of the Report (ASME Review Panel Members only)

5:00 p.m. Adjournment

Thursday, January 16, 2003

Radisson Riverfront Hotel Augusta

8:30 a.m. Writing of the Report (ASME Review Panel Members only)

5:00 p.m. Adjournment

Friday, January 17, 2003

Radisson Riverfront Hotel Augusta

8:00 a.m. Closeout Session

9:00 a.m. Adjournment

Executive Summary

The objective of the Nuclear Safety Program at Savannah River Site (SRS) is to manage SRS in a manner which ensures there is no undue risk of nuclear and process accidents that could adversely affect the health or safety of employees, visitors, members of the general public, or the environment. For all activities, the continued assurance of the capability and capacity for safe operations remains paramount to protect facilities and the environment from unacceptable risks. These programs apply to all personnel engaged in nuclear and process-related activities, and shall be in effect during design; construction; normal and abnormal operations; maintenance; modifications; surveillance; transition; deactivation; decontamination; and decommissioning. The intent is not only to prevent accidents, but to make provisions to limit consequences should accidents occur.

The Nuclear Safety Program is well-established and mature. Its functions are defined in a number of documents, and they are assigned to organizations having the appropriate authority and accountability. A recently-revised organizational chart illustrates the consolidation of all nuclear safety management responsibilities at SRS under a single manager, which is a positive step. The organizational interfaces could be better understood if there were a single document.

The Project Team is working on the Safety Analysis Documentation Improvement Strategy (SADIS) to streamline and better integrate the SRS procedures.

Training for newly-assigned engineers for the High-Level Waste Division systematically develops the required nuclear safety-related work process knowledge and skills. However, this high-quality program is not uniformly implemented across the site. The following training issues need to be addressed on a site-wide basis: lack of depth in technical disciplines and/or system specifics; facilities-specific training; continuing training program; lack of training program standardization; and course and/or instructor availability.

The application of the SRS Functional Classification process properly classifies the equipment; however, overclassification of some systems is possible.

The Nuclear Safety Program Performance Indicators are meaningful to the SRS management for identifying the number of significant nuclear safety issues and the corresponding status of backlog resolution. However, they are not sufficiently comprehensive. Additional performance indicators, tracked by the Project Team, could possibly be used to evaluate nuclear safety performance.

The Nuclear Safety Program Performance Indicators are not sufficiently predictive of nuclear safety problems, because they do not adequately account for the potential cumulative effect of "minor incidents" (including radiological and industrial safety).

The decision for entry of the New Information and Potential Inadequacy in Safety Analysis into the Unreviewed Safety Question (USQ) Process is "knowledge"-based, and therefore involves subjectiveness. For New Information and Potential Inadequacy in Safety Analysis evaluation, it is not necessary to rely on numerical threshold values.

There is no specific measurement of the adequacy of the communication of the USQ Process to field engineers. However, evidence of adequacy can be inferred from the results of the assessments performed by the Quality Assurance Department, the Facility Evaluation Boards, and the facilities.

RECOMMENDATIONS

Based on a careful assessment of the information presented to the Review Panel (RP) and the findings developed in response to the review criteria, the RP provides the following recommendations:

1. The Project Team should develop a single document that identifies all the elements of the Nuclear Safety Program.
2. In the near future, the Project Team should evaluate the effectiveness of the new organization that consolidates all nuclear safety management responsibilities.

3. The Project Team should complete the Safety Analysis Documentation Improvement Strategy (SADIS) in a timely manner.
4. The personnel performance evaluation system should be reviewed to reflect the alignment of nuclear safety responsibilities to individual accountability.
5. The Project Team should review the program for Washington Safety Management Solutions (WSMS) safety analysts assigned to Savannah River Site tasks to ensure, as a minimum, training in: the methodologies (including quantitative human reliability) used in safety assessment; and the facilities being analyzed.
6. The Project Team should address in a timely manner the following training issues on a site-wide basis: lack of depth in technical disciplines and/or system specifics; facilities-specific (e.g., process, safety basis) training; continuing training program; lack of training program standardization; and course and/or instructor availability.
7. The Nuclear Safety Program Performance Indicators should be normalized to account for the major facilities status during the measurement period.
8. The Nuclear Safety Metrics should reflect the full range of potential indicators that can be used for assessing the Nuclear Safety Program performance.
9. The Project Team should consider accounting for the potential cumulative effect of "minor incidents" (including radiological and industrial safety events) into the Nuclear Safety Program Performance Indicators.
10. The Project Team should complete in a timely manner its computerized Unreviewed Safety Question (USQ) Process desktop.

Peer Review Process

INTRODUCTION

There is consensus within the technical community on the definition, process, and key criteria for the acceptability of peer review. Peer review consists of a critical evaluation of a topic by individuals who—by virtue of their education, experience, and acquired knowledge—are qualified to be peers of an investigator engaged in a study. A peer is an individual who is able to perform the project, or the segment of the project that is being reviewed, with little or no additional training or learning.

Recognizing that peer review constitutes the core of acceptability of scientific and engineering information, virtually all professional societies of scientists and engineers have instituted formal procedures for peer review for their activities. The American Society of Mechanical Engineers (ASME), also known as ASME International, has over a century of experience in peer review. Consistent with its mission and tradition, ASME, in cooperation with the Institute for Regulatory Science (RSI), has established a peer review program devoted to the review of activities of various government agencies (ASME 2000a, RSI 2000). The reports of the peer reviews resulting from this program have been published (ASME 1997, 1998, 1999, 2000b, 2001a, 2001b, 2001c, 2002a, 2002b, 2002c).

PEER REVIEW PROCESS

The structure of the peer review process established by the ASME/RSI team consists of a tiered system. For each specific area, the entire process is overseen by a Peer Review Committee for Energy and the Environment (PRCEE). The review of specific topics is performed by Review Panels (RPs).

Peer Review Committee for Energy and the Environment

The PRCEE is a standing committee of ASME formed to oversee peer review for one particular program in an agency. Its members are chosen on the basis of their education, experience, peer recognition, and contribution to their respective areas of competency. An attempt is made to ensure that all needed technical competencies and diversity of technical views are represented in the PRCEE. The members of the PRCEE must be approved by the Board on Research and Technology Development of the Council on Engineering of ASME. The PRCEE includes an Executive Panel (EP) that is responsible for the day-to-day operations of the PRCEE. Except for the EP, membership in ASME is not required for appointment to the PRCEE. As the overseer of the entire peer review process, the PRCEE enforces all relevant ASME policies, including compliance with professional and ethical requirements. A key function of the PRCEE is the approval of the appointment of members of RPs for a specific project.

Review Panels

The review of a project, a document, a technology, or a program is performed by an RP consisting of a small group of highly-knowledgeable individuals. Upon the completion of their task, the RPs are disbanded. The selection of reviewers is based on the competencies required for the specific review assignment. The number of individuals in an RP depends upon the complexity of the subject to be reviewed. The selection of a reviewer is based on the totality of that individual's qualifications. However, there are several generally-recognized and fundamental criteria for assessing qualifications of a reviewer. These are as follows:

1. **Education:** A minimum of a B.S. degree, preferably an advanced degree in an engineering or scientific field, is required for any peer reviewer.

2. **Experience:** In addition to education, the reviewer must have significant experience in the area that is being reviewed.

3. **Peer recognition:** Election to an office of a professional society, serving on technical committees of scholarly organizations, and similar activities are considered to be a demonstration of peer recognition.

4. **Contributions to the profession:** Contributions to the profession may be demonstrated by publications in peer-reviewed journals. In addition, patents, presentations at meetings where the papers were peer-reviewed, and similar activities are considered to be contributions to the profession.

5. Conflict of Interest: One of the most complex and contested issues in peer review is a set of subjects collectively called conflict of interest. The ideal reviewer is an individual who is intimately familiar with the subject and yet has no monetary interest in it. Despite this apparent difficulty, the ASME and similar organizations have successfully performed peer review without having a real or apparent conflict of interest. The guiding principle for conflict of interest is as follows: *Those who have a stake in the outcome of the review may not act as a reviewer or participate in the selection of reviewers.*

Due to the multidisciplinary nature of many projects reviewed by the ASME/RSI team, rapid identification of qualified peer reviewers and their availability to participate in the review process are key ingredients for a successful program. The process used for the identification of reviewers is multifaceted. The Administrative Manager of the Peer Review Program receives recommendations from sources within ASME; previous members of the RP; sister societies; other organizations and individuals; the U.S. Department of Energy (DOE); DOE contractors; and others. However, the selection of peer reviewers is based entirely on criteria identified by ASME. The details of various aspects of peer review, including conflict of interest, can be found in the ASME *Manual for Peer Review* (ASME 2000a) and the associated procedures (RSI 2000).

COOPERATION WITH OTHER PROFESSIONAL SOCIETIES

The ASME is a large professional engineering society having in excess of 125,000 members. Although the predominant discipline of the members is mechanical engineering, there are members who—by virtue of their education, training, or experience—are competent in other disciplines. The Council on Engineering includes divisions ranging from classical mechanical engineering (design, heat transfer, and power) to solar engineering; environmental engineering; and safety and risk analysis. Despite the diverse competency within ASME, it is recognized that on occasion it will become necessary to peer review activities which include disciplines that are outside the areas of competency of ASME and its members. These disciplines may include geology, hydrology, toxicology, and ecology. Consequently, ASME has reached formal and informal agreements with its sister societies to identify qualified reviewers in areas outside of those covered by the membership of ASME.

PERFORMING ORGANIZATIONS

The Center for Research and Technology Development of ASME manages a number of scientific and engineering activities, including peer reviews. Because of ASME's conscious effort to maintain a small in-house staff, it relies upon other organizations to provide detailed project management services in its research, development, and similar activities. Accordingly, ASME and RSI joined forces in a collaborative effort to perform the peer review for the U.S. Department of Energy. While the ASME staff in Washington, DC provides the staff support for the PRCEE, the detailed management and staff support for the RPs is provided by RSI.

American Society of Mechanical Engineers

As one of the largest professional engineering societies, ASME has a long and distinguished history. Its activities are carried out primarily by members who volunteer their time in support of engineering and scientific advancement. For obvious reasons, ASME also has a paid staff to manage the day-to-day operations of such a large professional society. ASME has a detailed structure for its operation, consisting of councils, boards, divisions, and committees. The Council on Engineering has 38 divisions, including: Environmental Engineering; Solid Waste Processing; Nuclear Engineering; Safety Engineering; and Risk Analysis. The Council on Codes and Standards develops ASME codes and standards that are the backbone of many industries—including power production—worldwide. The Council on Codes and Standards is also responsible for the development of standards for activities such as certification of incinerator operators. The ASME was a founding member of the American Association of Engineering Societies and a founding member of the American National Standards Institute.

Institute for Regulatory Science

RSI is a not-for-profit organization chartered under section 501(c)3 of the Internal Revenue Service. It is dedicated to the idea that societal decisions must be based on the best available scientific and engineering information. According to the RSI mission statement, peer review is the foundation of the best available scientific and engineering information. Consequently, RSI has promoted peer review within government and industry as the single most important measure of reliability of scientific and engineering information. In its activities, RSI seeks the cooperation of scholarly organizations. Historically, a large number of RSI activities have been performed in cooperation with ASME. RSI is located in the Washington, DC metropolitan area.

Project Summary

INTRODUCTION

The Savannah River Site (SRS) occupies approximately 310 square miles (800 km^2), and is located in the southeastern coastal area of the United States in the state of South Carolina. It is bordered to the west by the Savannah River and Georgia. The closest cities to SRS are Augusta, GA; Savannah, GA; Columbia, SC; Greenville, SC; and Charleston, SC. Established in 1950 by the Atomic Energy Commission, the predecessor of the U.S. Department of Energy (DOE), SRS has been involved in a number of activities, notably the production of tritium and plutonium in support of the nuclear stockpile. Table 1 contains the summary of events (SRS 2002) leading to the 50th Anniversary of SRS on November 28, 2000.

Table 1. Historical highlights at Savannah River Site.

Year	Highlight
1950	E. I. duPont de Nemours and Company is asked by the Atomic Energy Commission to design, construct, and manage the Savannah River Plant. An area for the site is chosen.
1951	Savannah River Ecology Laboratory begins ecological studies of SRS plants and animals. Construction begins at the site.
1952	Production of heavy water for site reactors begins in Heavy Water Rework Facility.
1953	R-Reactor, the first production reactor, goes critical.
1954	P-Reactor, L-Reactor, and K-Reactor go critical. The first irradiated fuel is discharged. F-Canyon, a chemical separation facility, begins radioactive operations.
1955	C-Reactor goes critical. The first plutonium shipment leaves the site. H-Canyon, a chemical separation facility, begins radioactive operations.
1956	Construction of the basic plant is complete.
1963	Receiving Basin for Off-Site Fuels receives first shipment of off-site spent nuclear fuel.
1964	R-Reactor is shut down.
1968	L-Reactor is shut down for upgrades.
1971	K-Reactor becomes the first production reactor automatically controlled by computer.
1972	The site is designated as the first National Environmental Research Park.
1981	SRS begins environmental cleanup program. M-Area Settling Basin cleanup begins under the Resource Conservation and Recovery Act.
1982	Heavy Water Rework Facility is closed.
1983	Ground is broken for construction of the Defense Waste Processing Facility. Wackenhut Services Incorporated begins providing security support services at SRS.
1985	HB-Line begins producing plutonium-238 for NASA's deep-space exploration program. L-Reactor restarted and C-Reactor are shut down. A full-scale groundwater remediation system is constructed in M-Area.
1986	Construction of Saltstone begins. Construction of the Replacement Tritium Facility begins.
1987	DuPont notifies DOE that it will not continue to operate and manage the site.
1988	K-, L-, and P-Reactors are shut down. Effluent Treatment Facility begins operations to treat low-level radioactive wastewater from F- and H- Area Separations facilities.
1989	The site is officially included on the National Priority List and becomes regulated by the U.S. Environmental Protection Agency (EPA). Westinghouse Savannah River Company assumes management and operation of site facilities.

Table 1. (cont'd)

Year	Highlight
1990	Construction of a cooling tower for K-Reactor begins. Saltstone begins operations.
1991	Mixed Waste Management Facility is the first SRS facility closed and certified under the provisions of the Resource Conservation and Recovery Act. L-Reactor shut down. M-Area Settling Basin closure completed. Cold War ended and production of nuclear materials for weapons use at SRS stopped.
1992	K-reactor is operated briefly for last time and connected to cooling tower. Secretary of Energy announces phaseout of all uranium processing. Non-radioactive operations begins at the Replacement Tritium Facility.
1993	K-Reactor placed in cold-standby condition as Nation's tritium source. Non-radioactive test runs of the Defense Waste Processing Facility begin. Construction begins on Consolidated Incineration Facility. Tritium introduced into the Replacement Tritium Facility and radioactive operations begin. Workforce Transition and Community Assistance begins at SRS.
1994	SRS Citizens Advisory Board is established.
1996	The Defense Waste Processing Facility introduces radioactive material into the vitrification process. K-Reactor placed in shutdown condition. F-Canyon restarts and begins stabilizing nuclear materials at SRS.
1997	First high-level radioactive waste tanks closed.
1998	SRS is selected as the preferred site of three new plutonium missions.
2000	K-Reactor building is converted to K-Area Materials Storage Facility. Record of Decision is announced, selecting SRS as the site of new plutonium missions: MOX Fuel Fabrication Facility; Pit Disassembly and Conversion Facility; Plutonium Immobilization Facility. Westinghouse Savannah River Company (WSRC) Team earns the U.S. Department of Energy's top safety performance and program honor: Star Status. Defense Waste Processing Facility pours its 1,000th canister of glassified radioactive waste.
2001	Safety milestone: WSRC Operations employees worked ten-million hours without an injury resulting in time away from work. First shipment of transuranic waste is shipped to DOE's Waste Isolation Pilot Project in New Mexico. Environmental cleanup mark surpassed with the removal of industrial solvents from over four-billion gallons of groundwater. The Defense Waste Processing Facility takes the lead worldwide with the pouring of four million pounds of environmentally acceptable glassified waste. Over $44 million is saved in 2001 through the SRS employee suggestion program. SRS employee responses to the 2001 National Safety Council's safety perception survey say the site's safety culture is one of the best in the nation.
2002	After 50 years of service to the nation, the F-Canyon and FB Line facilities at SRS complete their last production run to process legacy materials.

The Savannah River Operations Office (SROO 2002) focuses on three mission areas in overseeing the management and operations contractor and other contractors at SRS, a key DOE facility, associated with products and services essential to achieving the DOE's goals:

1. Nuclear Weapons Stockpile Stewardship. Stewardship is the responsibility for the careful use of money, time, talents, and other resources, especially with respect to the principles and needs of a community. Nuclear Weapons Stockpile Stewardship emphasizes science-based

maintenance of the nuclear weapons stockpile. SRS supports the stockpile by ensuring the safe and reliable recycle, delivery, and management of tritium resources; by contributing to the stockpile surveillance program; and by its ability to assist in the development of alternatives for large-scale pit production capability, if required.

2. Nuclear Materials Stewardship. Nuclear Materials Stewardship is the management of excess nuclear materials, including stabilization, storage, transportation, and disposition to support nuclear nonproliferation initiatives. Primary nuclear materials in this program include components from dismantled weapons; residues from weapons processing activities; spent nuclear fuel; and other legacy materials.

3. Environmental Stewardship. Environmental Stewardship involves management including treatment, and disposal of radioactive and other wastes resulting from past, present, and future operations. This stewardship includes pollution prevention and restoration of the environment impacted by site operations. Environmental Stewardship also encompasses the site's extensive natural and cultural resources.

Corporate Management addresses the successful accomplishment of the three stewardship missions through five fundamental principles or Focus Areas: Safety and Security; Technical Capability and Performance; Community, State, and Regulator Relationships; Cost-Effectiveness; and Corporate Perspective. These Focus Areas crosscut all site functions, defining how work is done at SRS, with an emphasis on achieving results and continually improving performance.

MANAGEMENT POLICIES

Consistent with the three stewardship principles, the Westinghouse Savannah River Company (WSRC) has developed the *Management Policy* 4.5 (WSRC 2002a) which requires the Savannah River Site (SRS) to be managed in a manner so that there is no undue risk of events that could adversely affect the health or safety of employees, visitors, members of the general public, or the environment. For all activities, the continued assurance of the capability and capacity for safe operations remains paramount to protect facilities and the environment from unacceptable risks. The scope of this policy is defined broadly to cover environmental, safety, and health topical areas. The intent is not only to prevent accidents, but to make provisions to limit consequences should accidents occur.

An *Integrated Safety Management Program* is used to implement this policy. According to the implementation program, this policy shall apply to all personnel engaged in nuclear and process related activities for WSRC, and shall be in effect during design; construction; normal and abnormal operations; maintenance; modifications; surveillance; transition; deactivation; decontamination; and decommissioning. Scope of Work for nuclear and process facilities shall be clearly defined in support of the site missions. Activities within the Scope of Work shall be prioritized and resources properly allocated such that work can be performed safely. Each nuclear facility shall have a Safety Basis and, when directed by the U.S. Department of Energy (DOE), an Authorization Agreement. The site Criticality Safety Program shall be applicable to any process, structure, or system that requires the control of one or more parameters for Criticality Safety purposes. Revisions to the site Criticality Safety Program shall have the approval of the Nuclear Criticality Safety Review Committee. The information on any nuclear facility formally transmitted to DOE shall be complete (i.e., sufficient information considered necessary to understand an issue) and accurate. The verification of any information must include a consideration of consistency with the Safety Basis, Authorization Agreement and supporting documentation, and the as-built facility. Administrative and procedural controls shall delineate clear lines of responsibility and methods for safe operation under normal and emergency conditions. Activities related to nuclear and process facilities shall be accomplished in accordance with the applicable requirements agreed upon with the DOE and included in the WSRC Standards/Requirements Identification Documents (S/RIDs); federal and state requirements; appropriate consensus codes and standards; and WSRC policies and procedures. These activities shall also address appropriate commercial industry nuclear and process safety practices, if determined to be appropriate and practical for SRS. Where established standards are inadequate or unavailable, WSRC shall develop suitable standards for the establishment of a defined, and agreed upon, basis for conducting and assessing nuclear and process operations. Nuclear and process related activities shall be conducted according to detailed written documents and procedures approved in advance that assure adequate consideration of health and safety factors.

The same policy requires that Operational Readiness Reviews (ORR) for nuclear facilities shall be performed prior to initial startup or prior to restart following shutdown and major modifications. WSRC shall have a program that proactively identifies deficiencies, determines root causes, and establishes and implements timely corrective actions to prevent recurrence of the specific event and addresses the generic implications. The program shall include provisions for implementing and maintaining the commitments and configurations resulting from the corrective actions. WSRC shall also proactively incorporate, as appropriate, lessons-learned or state-of-the-art methodology and technology as a method of applying continuous improvement. Continuous improvement and assurance of nuclear and process safety shall be achieved through the implementation of the Integrated Safety Management Program, which includes, but is not limited to: Independent Assessment by the Facility Evaluation Boards (FEB); Self-Assessment; Management Evaluation; Reviews; and Technical Support. This will assure WSRC itself, to DOE, and to the public that facility management and operations are safe and in accordance with applicable Laws, Regulations, Standards, guidance and the S/RID requirements.

Reviews are performed to assure accuracy and completeness of documents; authorization to start-up; or authorization to continue, e.g., ORR or QA inspections at specified hold points. All SRS nuclear and process related structures, programs, and operations shall be subject to independent assessment at all life-cycle stages.

WSRC Divisions/Departments shall be responsible for ensuring compliance with the policy, requirements, and references of this policy. This includes the following nuclear and process safety responsibilities:

1. Ensuring the protection of the public, the worker, and the environment
2. Ensuring that lines of authority and individual responsibilities for safety are clearly established, articulated, and understood
3. Ensuring that nuclear and process safety requirements are met
4. Developing and fostering technically competent personnel and technical standards necessary to achieve nuclear and process safety
5. Maintaining auditable training and qualification programs and operation, inspection, internal surveillance, and maintenance records
6. Allocating resources effectively to address nuclear and process safety and operational considerations
7. Using established nuclear and process safety goals to serve as aiming points for performance
8. Promoting a culture that continuously strives for nuclear and process safety enhancement
9. Complying with and resolving any shutdown or stop work actions by the operating staff, or directives by overview personnel
10. Establishing and maintaining a self-assessment program
11. Establishing a program that proactively identifies deficiencies, determines root causes, and establishes and implements timely corrective actions to prevent recurrence of the specific event and address the generic implications
12. Ensuring that information transmitted to DOE is complete and accurate

In addition to the responsibilities listed for all Divisions/Departments, the Environment, Safety, Health, and Quality Assurance Division is responsible for:

1. Preparing and maintaining this Management Policy
2. Performing vigorous Independent Assessment of activities and of the programs for managing these activities through the FEB and ORR to assure compliance with the S/RID requirements and assuring enhanced nuclear and process safety and reliability
3. Ordering suspension of nuclear, process facility or related operations and processes when it is determined that continued operation would jeopardize the health and safety of employees or the general public

4. Establishing and maintaining a list of SRS nuclear and radiological facilities
5. Performing activities and programs for managing these activities to ensure compliance with the S/RID requirements and ensuring enhanced nuclear and process safety and reliability
6. Providing Technical Support and reviews of progress toward established company safety, environmental, and quality performance goals in each program and activity area
7. Coordinating the following site activities, all of which are associated with nuclear and process safety: DOE Directives Administration Program, Lessons Learned Program, Root Cause Program, Site Item Reportability, and Issues Management Program

In addition to responsibilities listed for all Division/Departments, the ODs (i.e., the divisions responsible for managing nuclear and process facilities) are responsible for:

1. Providing training for their personnel to achieve the competency necessary for the assigned duties
2. Assuring the use of applicable safety codes and standards, including those referenced by, or contained in, the S/RID, and in the design and/or modification of nuclear and process facilities
3. Preparing and maintaining applicable design criteria, safety analyses, and other project documents and drawings affecting nuclear and process safety for appropriate review prior to implementation
4. Documenting and maintaining Safety Basis documents (including analyses) and Authorization Agreements
5. Documenting and maintaining the configuration of their facilities per configuration management program
6. Operating, maintaining, and ensuring the safety of their facilities in accordance with approved WSRC procedures, Safety Basis documents, Authorization Agreements, and operating and administrative controls
7. Reviewing their nuclear facility operations to ensure conformance with the safety envelope defined by the Safety Basis, Authorization Agreements, and supporting analyses and documents
8. Ensuring that design and construction of new facilities and renovation and modifications of existing facilities are performed in a safe, secure, cost-effective, environmentally sound, and quality manner

In addition to responsibilities listed for all Divisions, Departments, and Projects, the Engineering and Construction Division is responsible for providing engineering and technical support, when requested. The Site Chief Engineer is responsible for:

1. Establishing and maintaining a site-wide configuration management program
2. Establishing and maintaining a site-wide nuclear criticality safety program, including performing authorization basis criticality analyses
3. Establishing and maintaining a site-wide Safety Documentation Program for the development and maintenance of the Safety Basis, Authorization Agreement, and other safety documents

SAFETY ANALYSIS AND DOCUMENTATION ACTIVITIES

The *Facility Safety Document Manual* (WSRC 2001) describes requirements and guidelines that apply to Westinghouse Savannah River Company (WSRC) safety analysis and documentation activities. The term "WSRC" in this Manual includes all partners: Bechtel Savannah River Inc.; B&W Savannah River Company; British Nuclear Fuels, Limited Savannah River Corporation; and subcontractors. For the purpose of safety analysis and documentation, WSRC-11Q establishes uniform requirements for:

1. Determining Facility Hazard Categories (FHCs) for all facilities. The term "facility" is all inclusive and incorporates activities, projects, as well as buildings. Collocated and external safety support functions are included as part of the facility definition.

2. Preparing, reviewing, approving, and revising safety analyses for all facilities and operating modes. Such activities include identifying and controlling hazards, documenting results of safety analysis, and establishing and maintaining the safety basis.

3. Implementing U.S. Department of Energy (DOE) regulations (DOE 2001a) dealing with basic safety requirements and associated standards and guides.

A cornerstone of this Manual is placing facilities into one of seven FHCs for the purpose of determining necessary safety analysis and safety documentation requirements. The FHCs are based on radiological and hazardous material inventories. The complexity and depth of analyses and the level of detail in safety documents depend mostly on facility hazard, complexity, and lifetime. The hazard is measured in terms of FHC which, in turn, is based primarily on the quantities, locations, and forms of radiological and hazardous material inventories present. The FHC has a major impact on analysis detail, types of safety documents, and facility design criteria, and requirements and standards. The FHCs are:

1. HC-1, HC-2, HC-3 Nuclear Operations involving radioactive materials that present a significant hazard to workers or the public. This is based on a DOE definition. "Significant" was added to denote the boundary with Radiological. The methodology for categorizing facilities as Nuclear or Radiological is based on DOE-STD-1027-92 (DOE 1997) which allows credit for material form, quantity, and location. Nuclear Facilities are divided into the following Hazard Categories based on the radiological consequences of the safety basis inventory as follows:

   1.1 HC-1: potential for significant off-site consequences
   1.2 HC-2: potential for significant on-site consequences
   1.3 HC-3: potential for only significant localized consequences

2. Radiological Operations (also called Below HC-3) involving radioactive materials that do not present a significant hazard, but do involve quantities above a minimum level of concern. Radiological Facilities are Below HC-3 Nuclear Facilities with radiological hazards below DOE-STD-1027, but above a minimum U.S. Environmental Protection Agency (EPA) level of concern, i.e., Reportable Quantities (RQ)s, defined in the EPA regulations 40 CFR 302.4 (EPA 1999).

3. Chemical Operations involving chemical hazards in forms or quantities beyond common industrial settings. Chemical hazards are subdivided further into High Hazard Chemical and Low Hazard Chemical for flexibility in implementing requirements. Chemical Facilities are divided into the following categories:

   3.1 Those inventories that require an OSHA Process Safety Management (PSM) program or an EPA's Risk Management Program (RMP) (designated as High Hazard Chemical Facilities); and
   3.2 Those inventories that do not require PSM/RMP (designated as Low Hazard Chemical Facilities).

4. Other Industrial Operations involving hazardous materials routinely encountered in public or common industrial settings. Other Industrial Facilities are those facilities with hazards below the EPA minimum level of concern for both radiological and other hazardous material.

A facility may be divided into individual segments in order to assign segment hazard categories based on segment inventory, provided those segments can be considered independent. Thus, segmentation is permitted only if the hazardous material in one segment cannot interact with hazardous material in other segments and safety systems do not cross over. Segmentation provides an opportunity to apply a graded approach for facility safety documentation.

FHCs allow for a graded approach in developing safety documents. For example, an HC-1, HC-2, HC-3 Nuclear Facility will require safety documents with more formality and detail than will a Radiological Facility. Also a High Hazard Chemical Facility will require safety documents with more formality and detail than will a Low Hazard Chemical Facility. Other Industrial Facilities may not need safety documents except for the Administrative Limits (ALs). Hazards are identified from material inventories and energy sources and are recorded by amount, physical and chemical characteristics, and location. Common industrial material is screened out from further analysis. Finally, inventories are selected as the basis for safety analysis conclusions and controls. The process must be comprehensive so that important hazards are not overlooked and potential interactions are considered. The inventories that represent the safety basis must be conservative and must consider operating fluctuations because they are the basis for operating restrictions and administrative controls. For Other Industrial Facilities, hazard identification should suffice as the only safety analysis required. Common industrial hazards are not analyzed further because they are addressed by federal regulatory compliance, general industrial safety, or "routine" hazards. Consensus industrial standards for handling and treating common industrial and routine hazards are considered adequate control. Examples of hazards that may be excluded are: material in forms or quantities commonly used or encountered and accepted by the public; chemicals purchased in quantities that are adequately controlled by industrial standards; chemicals in "laboratory quantities"; chemical inventories that place a facility in the Other Industrial category; chemicals that are inert, do not interact with other chemicals, pose no health hazard, or are in a non-dispersible form. Radioactive material used as sealed sources is added to the safety basis inventory only if the sealed source is not procured, manufactured, or handled in accordance with industry standards.

There are several types of safety documents, some of which are included in a safety basis (SB):

1. Documented Safety Analysis (DSA). A DSA is a documented analysis of the extent to which a HC-1, HC-2, HC-3 Nuclear Facility can be operated safely with respect to workers, the public, and the environment, including a description of the conditions, safe boundaries, and hazard controls that provide the basis for ensuring safety. A DSA normally consists of either a Safety Analysis Report (SAR) or a Basis for Interim Operation (BIO). A "graded approach" to DSAs is provided via the "Safe Harbor" approach. Converting existing safety documents to a "Safe Harbor" method may meet Rule requirements provided these conversions are approved by DOE by means of an SER. Such conversions should be handled on a case-by-case basis. The "Safe Harbor" approach is explained in detail in Procedure 1.01 of the *Facility Safety Document Manual* (WSRC 2001). The DSA is approved by DOE through issuance of an SER.

2. Hazard Assessment Document (HAD). The HAD presents the results of hazard identification and categorization. Thus, the HAD includes the inventory and FHC along with associated facility segmentation. If hazard analyses information or conclusions are presented in other documents, a stand-alone HAD may not be necessary. Because HADs generally do not contain information related to adequacy of preventive or mitigative measures, a stand-alone HAD may not satisfy Auditable Safety Analysis (ASA) requirements.

3. Health and Safety Plan (HASP). As required by Occupational Safety and Health Administration (OSHA) (2002a) regulation, HASP (WSRC 2003) is prepared for activities involving hazardous waste remediation or treatment, storage, or disposal facilities. The site-specific HASP can provide essential hazard analysis information so that such information need not be repeated in other documents. HASPs can play a role as an ASA. In some cases, the OSHA regulation allows the HASP to play a role in DSAs. In those special cases, the HASP must satisfy DSA Rule requirements.

4. Process Hazard Analysis (PrHA). PrHAs are required for any processes involving hazardous chemicals at or above the specified

threshold quantities to satisfy the OSHA (2002b) regulation 29 CFR 1910.119. PrHAs are also required by EPA (2002) regulation 40 CFR 68, known as Risk Management Program. The PSM program consists of 14 elements and PrHA is one of the key elements in the PSM program. The PrHA is also one of the key elements in the RMP. The RMP rule requires that owner/operator of a stationary source of a regulated substance to submit a single Risk Management Plan (RMPlan) which includes a summary of PrHA. Both PrHA and RMPlan can provide essential hazard evaluation information so that such information need not to be repeated in other documents. PrHAs and RMPlan can play a role for High Hazard Chemical facilities as the ASA. It should be noted that PSM-PrHAs focus on worker protection while RMP-PrHAs focus on protection of the public and environment off-site. Therefore, these PrHAs may require some augmentation to become safety basis ASA documents.

5. Preliminary Hazard Analysis (PHA). The PHA is developed as support for HC-1, 2, 3 Nuclear Facilities, SARs, or BIOs. Although separate PHAs are not expected for Non-nuclear Chemical facilities, any existing PHAs can play a role as the ASA. PHAs meet the need of hazard analysis/documentation for worker protection against nuclear/chemical process hazards for those facilities containing such hazards. There has been confusion regarding various terms and acronyms for "hazard analysis" due to differences of terminology between DOE guidelines and chemical industry guidelines. The DOE terminology is adopted, i.e., PrHA means Process Hazard Analysis per OSHA-PSM or EPA-RMP rules, and PHA means Preliminary Hazard Analysis per DOE-STD-3009 (DOE 2000a).

6. Process Hazard Review (PHR). The PHR program predates the OSHA-PSM or EPA-RMP rules, and they are conducted at the discretion of the facility on many operations regardless of chemical inventories. Guidelines for conducting PHRs and documentation are established in *SRS Process Safety Management Manual* (WSRC 2002e). PHRs can support Nuclear Facility SBs. However, PHRs are not SB documents. The PHR can provide essential hazard evaluation information so that such information need not to be repeated in other documents. PHRs can play a role as the ASA. The PHR program meets the need of hazard analysis/documentation for worker

protection against nuclear/chemical process hazards for those facilities containing such hazards. The content of a PHR does not cover all the elements required in a PrHA, hence a PHR may not replace a PrHA without some additional information.

7. Emergency Preparedness Hazard Assessment (EPHA). EPHAs document a special hazard assessment that satisfies emergency management program requirements from DOE Order 5500.3A, *Planning and Preparedness for Operational Emergencies* (DOE 1992) by complying with the *Standard for Development and Maintenance of Hazard Assessment* (WSRC 2002d). This assessment is used to determine the extent and scope of emergency planning and preparedness activities. EPHA information may be useful for safety analyses.

8. Fire Hazard Analysis (FHA). FHAs document a special fire hazard assessment. FHAs are required for Nuclear Facilities, all new facilities, and any major modification of existing facilities. Results from a FHA can be incorporated into a SAR for a Nuclear Facility.

9. Safety Analysis Report (SAR). The SAR is the primary SB document for a Nuclear Facility. The SAR presents the results of hazard evaluation and accident analyses, and supporting information to demonstrate the facility as designed and constructed can be operated without undue risk to workers, the public, and the environment. The SAR describes site, facility, and operation characteristics; defines the safety basis; presents hazard and accident analyses; identifies SCSSCs, SSSSCs, and operating limits derived from the analyses; describes administrative programs that support safe operation (e.g., emergency preparedness, fire protection, criticality safety, etc.). Supporting and generic site information can be referenced. Generic site information may appear in a single Generic SAR (GSAR). A special SAR, called a Safety Analysis Report for Packaging (SARP), addresses transportation. SARPs are covered by the *Transportation Safety Manual* (WSRC 2002f).

10. Basis for Interim Operation (BIO). BIO can represent the DSA for short-life operations or facilities in transition from operating missions.

BIO requirements are based on DOE-STD-3011-94, *Guidance for Preparation of DOE 5480.22 and DOE 5480.23 Implementation Plans* (DOE 1991).

11. Technical Safety Requirements (TSRs)/Operational Safety Requirements (OSRs). Nuclear Facilities are operated in accordance with TSRs, covered by the Rule. TSRs replace OSRs, and their bases, and the management or administrative controls required to assure safe operation. Thus, TSRs/OSRs are an integral part of an SB. TSRs for HC-3 facilities may consist of only inventory control and commitments to safety programs.

12. Justification for Continued Operations (JCOs). The JCO (or equivalent document such as Safety Evaluation) has been used both to document the current safety envelope for facilities that do not conform to current safety documentation requirements and to provide an interim basis if an extended analysis is needed during the Unreviewed Safety Question (USQ) process. The former objective is similar to the objectives of the BIO discussed above. In this case the JCO is called a Technical JCO or TJCO. If a TJCO or JCO is used as part of an HC-1, HC-2, HC-3 Nuclear Facility safety basis, then it must satisfy the requirements of the Rule.

13. Auditable Safety Analysis (ASA) and ALs. ASA documentation is a graded approach to providing safety analysis documents for Chemical and Radiological Facilities. ASA objectives are to provide systematic identification of hazards and to describe and analyze the adequacy of measures to eliminate or control those hazards. These objectives can be satisfied by other documents prepared in the course of performing a safety analysis. But, if existing documentation does not provide sufficient information, other safety documents should be prepared containing the additional information. For High Hazard Chemical Facilities, the ASA objectives may be covered by PrHA required by OSHA-PSM or EPA-RMP rules. For those facilities or activities under Resource Conservation and Recovery Act (RCRA), ASA objectives may be met by HASPs required by OSHA. ALs are a graded approach to providing operating limits for Non-nuclear Facilities or Other Industrial Facilities. Procedure 1.03 of the *Facility Safety Document Manual* (WSRC 2001) contains the minimum

ALs for Other Industrial Facilities. ALs preserve the Facility Hazard Category, and important safety basis conclusions and assumptions. If other safety documents provide enough information to serve the same function as ALs, then separate ALs are not needed. If existing documents do not provide sufficient information, ALs will protect the assumptions and conditions significant to the Facility Hazard Category.

Operations are conducted at WSRC by procedure. After a safety document is added to the safety basis, then, as needed, procedures are revised or added to incorporate the new requirements, assumptions, equipment, or limits. Thus, there is a link between the safety basis and operating procedures. Because of the importance of this link, it is also important that operating procedures that can affect the safety basis are identified and are subject to safety basis control requirements. Developing and maintaining linking documents (LDs) are optional and are not required for any WSRC facilities.

Authorization Agreements (AAs) set forth the bases on which DOE authorizes WSRC to perform operations at selected facilities. The AA, initiated by DOE-SR, establishes and documents in one concise document the terms and conditions binding on WSRC for a particular scope of operation at the selected facilities in a manner that preserves DOE's bases for authorizing facility operations. Those bases typically include the safety, environmental, regulatory, readiness, and competency bases.

Safety document control is one part of the process by which changes to facilities are authorized, and revisions to safety documents are made. Thus, control consists of authorization of proposed activities and maintenance of safety documents.

Authorization of changes differs depending on the FHC:

1. HC-1, HC-2, HC-3 Nuclear Facilities. The review to determine whether changes to Nuclear Facility configuration, procedures, or management systems require DOE authorization prior to implementation is called the Unreviewed Safety Questions (USQ) process. Such changes are called Proposed Activities (PAs). Basically, PAs are reviewed against the SB, in accordance with Procedure 1.05 of

the *Facility Safety Document Manual* (WSRC 2001) to determine if DOE must authorize the PA. In addition, there are other supporting safety analyses or documents that are used to make this USQ determination. These supporting safety analyses or documents together with existing SB documents are considered as a baseline for USQ Evaluation (USQE). PAs that do not require DOE approval prior to implementation are considered to be "within the SB" regardless of whether revisions to SB documents are required. Thus, they are considered to be pre-authorized by DOE. PAs can include changes to equipment, procedures, processes, or administrative controls. Discoveries of inadequacies in the SB are also reviewed. First, the discovery is screened to determine if it is a Potential Inadequacy in the Safety Analysis (PISA). If the discovery is a PISA, it is evaluated for a USQ. Screening prevents unnecessary USQ analysis for minor deficiencies. Further details on PISAs are in Procedure 1.04 of the *Facility Safety Document Manual* (WSRC 2001). Any changes to the TSRs/OSRs require DOE approval prior to implementation.

2. Radiological and Chemical Facilities. For Radiological and Chemical Facilities, the safety documents are authorized by WSRC. Although there are no USQ-type requirements by DOE Orders, it is a "good engineering practice" to properly manage changes related to facility configuration and operation. Facility-specific procedures should be established for the management of changes. Changes to facility configuration, procedures, or management systems are reviewed to determine if the changes need to be authorized prior to implementation. Existing procedures should be used to the maximum extent possible. Changes that increase or decrease the FHC are not implemented prior to satisfying safety document and other ES&H requirements for the new category. If the new category requires DOE authorization, (i.e., HC-1, HC-2, HC-3 Nuclear), then DOE authorization must be obtained prior to implementing the change. If the new category decreases from one requiring DOE authorization to one where DOE authorization is not required, DOE concurrence with the new category must be obtained prior to implementing the change. Changes that do not increase or decrease the FHC can be implemented before updating safety basis documents.

3. Other Industrial Facilities. For Other Industrial Facilities, the safety documents are authorized by WSRC. The ALs described in Procedure 1.03 of the *Facility Safety Document Manual* (WSRC 2001) are the only safety documents required for Other Industrial Facilities and hazardous materials inventory should be closely monitored. Special attention on inventory control should be paid by those Other Industrial Facilities whose inventories fluctuate significantly or challenge the RQs. Changes that increase the FHC are not implemented prior to satisfying safety document and other ES&H requirements for the new category. If the new category requires DOE authorization (i.e., HC-1, 2, 3 Nuclear), then DOE authorization must be obtained prior to implementing the change. Changes that do not increase the FHC can be implemented before updating safety documents.

Maintaining safety documents current also depends on the FHC:

1. HC-1, HC-2, HC-3 Nuclear Facilities. The USQ process is also used to determine how SB documents should be maintained. After SB documents are determined to require revision and the changes to operation are authorized, then the SB documents can be updated by incorporating the USQE and attached analysis/documents into the SB documents. In addition to maintenance for changes, documented safety analyses (SARs and BIOs) and TSRs shall be reviewed annually and updated and submitted to DOE-SR. The submittal of annual reviews and updates should coincide with the submittal of annual USQ summary reports.

2. Radiological and Chemical Facilities. For Radiological and Chemical Facilities, safety document maintenance depends on whether the proposed change affects the FHC. For proposed changes that change the FHC, safety document requirements for the new category must be satisfied. If the new category requires safety documents that need DOE approval, then such approval must be obtained as part of the authorization process. Safety document requirements for lower FHCs may be satisfied by the existing safety documents for the higher category. Safety document changes are processed in accordance with Procedure 1.01 of the *Facility Safety Document Manual* (WSRC 2001) and in accordance with any controlling procedures for the particular documents identified as the ASA. ASA documents are

reviewed and updated annually, if necessary. The annual review should be conducted by personnel with appropriate knowledge of the facility operation and hazard analysis technique. Safety basis control requirements such as update frequency or review/approval authority for stand-in safety documents (e.g., HASP, PrHA) must satisfy both this Manual and other relevant manuals.

3. Other Industrial Facilities. Safety document maintenance for Other Industrial Facilities also depends on whether the proposed changes affect the FHC. For changes that increase the FHC, safety document requirements for the new category must be satisfied. If the new category safety documents need DOE approval, then such approval must be obtained as part of the authorization process. Safety document changes are processed according to Procedure 1.03 of the *Facility Safety Document Manual* (WSRC 2001).

The Disposition phases of a facility begin when DOE terminates facility operations and facilities are declared "surplus" or "excess". Whereas the SB for HC-1, 2, 3 Nuclear Facilities during the Operations portion of their life-cycles may remain relatively constant, facilities being dispositioned may need more frequent and significant SB changes. Typically, Disposition employs processes that differ significantly from the original mission, and Facility Hazard Categories (FHC) are reduced as a result of removing inventories. This dynamic process must be coordinated to maintain consistency between the safety basis and the actual status of a facility. As the FHC changes, opportunities exist to reduce a major portion of the facility requirements burden by modifying the safety basis and obtaining the appropriate level of approval. For example, when the FHC drops below HC-1, 2, 3 Nuclear, the SB for the facility may be retired by using the following process: First, the new FHC should be documented in a HAD. When the HAD is approved, facility management should formally request the responsible DOE-SR Assistant Manager to concur with the new FHC and to approve retiring the SB. Although DOE approval is not required for HADs, that document should be sent along with the formal transmittal to help support the request. After written permission is obtained from DOE, the SB may be retired. Until that permission is received, the SB remains active and all TSR and other SB commitments must be met. ASA documentation must be available to replace the SB documentation, unless the FHC is classified as Other Industrial.

POTENTIAL INADEQUACY IN SAFETY ANALYSIS DECLARATION PROCESS

Procedure 1.04 of the *Facility Safety Document Manual* (WSRC 2001) provides a mechanism for reviewing New Information (NI). Initially, information is reviewed to determine if it should be processed as NI or referred to other closure systems (e.g., nonconformance report, occurrence reporting, proposed activity). NI is information that might jeopardize the adequacy of the Safety Analysis that supports the Safety Basis for a Nuclear Facility. A Potential Inadequacy in the Safety Analysis (PISA) is one condition considered for Unreviewed Safety Question (USQ) implications. A systematic process is provided to properly identify and assess NI from multiple sources in a timely fashion to determine if a PISA should be declared.

Procedure 1.04 of the *Facility Safety Document Manual* (WSRC 2001) applies to personnel identifying and reviewing NI related to Nuclear Facilities subject to the requirements of 10 CFR 830.203 (DOE 2001a) as included in the *Standard/Requirements Identification Document* (WSRC 2002c). This procedure does not override any requirements for immediate corrective action or the use of methods outside of line management to voice a concern. It also does not address human resource concerns or strictly Industrial Safety concerns, as they could not render the Document Safety Analysis (DSA) indeterminate for operation, components, or structures. Site-level NIs are processed in the same manner as Facility NIs.

The Facility Manager or designee is responsible for:

1. Declaring a PISA when identified
2. Imposing operating restrictions if warranted to place the facility in a safe condition, in accordance with Procedure 1.05 of the *Facility Safety Document Manual* (WSRC 2001)
3. Implementing required actions identified in Procedure 1.05 of the *Facility Safety Document Manual* (WSRC 2001) upon declaration of a PISA

The Originator is responsible for:

1. Initiating the review process, as soon as possible, by completion of the "Identification" block of the NI Processing Form and submitting the form to the Division NI Committee Member
2. Obtaining review and concurrence from the affected NI Committee Member for all NIs

Level 1 Managers are responsible for assigning their Division NI Committee Member. The Level 1 Managers responsible for control of site programs or services that may impact a DSA (e.g., road closures, controlled burns, WSI exercises, or changes to site-wide programs) are responsible for designating a Point-of-Contact to interface with the NI Committee.

The Point-of-Contact is responsible for:

1. Monitoring assigned site programs or services for changes that may impact a DSA
2. Informing the NI Committee of changes to site programs or services that may currently be relied upon in a DSA before the changes are implemented

The Site Chief Engineer assigns the New Information Committee Coordinator. The NI Committee Coordinator is responsible for:

1. Assuring all NIs entered in the NI database are presented to all NI Committee Members when further evaluation per this procedure is required
2. Providing monthly reports on all open NIs to the NI Committee
3. Forwarding the Originator a copy of the completed NI Processing Form
4. Establishing an electronic database that is accessible by the Committee Members and the U.S. Department of Energy-Screening Reviewer (DOE SR) to track NI to closure
5. Informally notifying DOE-SR of NIs that cannot be resolved within 10 calendar days from initiation

The NI Committee Members are responsible for:

1. Reviewing those planned temporary or permanent changes to site programs or services referred to the committee that could impact nuclear facility operations
2. Managing the program for the NI process described in this procedure
3. Reviewing NI for applicability within their respective Divisions
4. Assisting in resolution of technical issues as requested by Facility Management
5. Reviewing the status of all "open" NI on a routine basis to assure closure within a reasonable time
6. Reviewing for the appropriate Division Chief Engineer, on a case by case basis, the safety significance of a NI (e.g., where special attention is warranted or timely closure is in question)
7. Directing the Originator of concerns to other resolution systems when the PISA declaration process is not applicable

The NI Committee, as a minimum, is composed of one representative from each of the following:

1. Projects and Engineering and Construction (PE&C) Division (an individual familiar with the USQ process)
2. PE&C Division (an individual familiar with safety analysis and engineering)
3. Each Division Regulatory Programs Section
4. Safety Management Program Section, the Environment, Safety, Health, and Quality Assurance Division

Each member of the NI Committee shall appoint an alternate. All other Savannah River Site Divisions and other agencies (e.g., AID, SRFS, TSD, and Wackenhut) that control site programs or services with potential to impact a DSA assigns a "Point-of-Contact" to interface with the NI Committee. The Point-of-Contact is not a member of the NI Committee. Each Point-of-Contact, as a representative of site programs or services, shall identify to the NI Committee temporary or permanent changes to those programs, services, or activities (that have a potential for adverse impact) prior to implementation. Examples could include road closures, controlled burns, WSI exercises, or changes to site-wide programs. The NI Committee reviews a planned change to determine if it should be included in the NI Process. All Committee resolutions on the site program changes are documented in the committee meeting minutes. The NI Committee will act on all disagreements identified (e.g., adequacy of closure plan). If a consensus cannot be reached, the Coordinator will present, or will assign a Committee Member to present, the majority and minority positions at the next available Chief Engineers Meeting for resolution. All disagreements as identified in the agenda, raised by the Committee and resolutions of those disagreements will be documented in the meeting minutes.

The Regulatory Programs Section, or an organization assigned by management, is responsible for:

1. Assigning a NI Committee Member
2. Assigning a unique tracking number to a NI
3. Dispositioning NIs
4. Contacting the NI Committee Coordinator for assistance, if needed
5. Providing the NI Committee with closure plans to resolve open NI including a forecast for completion, as requested
6. Establishing a system to track the NI within a Division to closure
7. Maintaining NI Processing files
8. Providing the NI Committee Coordinator notice of initiation, disposition, and closure of NIs

The process of assessing NI for consideration as a PISA is divided into three steps. As shown in Fig. 1, these three steps are reflected on the NI Processing Form:

1. Identification. Upon identification of NI, the Originator completes the "Identification" block of the NI Processing Form and attaches applicable supporting information. The Originator provides the description of the NI, the source document (if applicable), and affected Divisions (if known) and signs the form. Identification of the Originator on the NI Processing Form is to ensure feedback is provided during and upon closure of the NI. The Originator, as soon as possible, obtains a review and concurrence from the NI Committee Member of the affected Divisions (if known) on the determination that the NI Process is appropriate for the information identified. The Committee Member signs the NI Processing Form and, if in concurrence with the Originator, forwards the form to the responsible

Regulatory Programs Section and the NI Committee Coordinator. When a specific Division has not been identified, the Committee Member forwards the NI form to the NI Committee Coordinator for distribution and evaluation. If the NI Committee Member does not concur that the NI Process is appropriate and the Originator disagrees with the action taken, then the Originator may contact the NI Committee Coordinator and request that the disagreement be addressed at the next NI Committee meeting. The Originator or someone the Originator designates as a representative shall be present. The actions of the Committee on the issue will be documented in the meeting minutes. The NI Committee Coordinator or an individual assigned by the Regulatory Program Manager logs the NI into the NI database, and forwards a copy to all NI Committee Members for evaluation. Each Committee Member reviews the New Information

1. NI # ________ NI PROCESSING FORM Date: ________

IDENTIFICATION (Originator)

2.a Brief Description of New Information (Complete block and sign below)

2.b Detailed Description of New Information (Attach additional sheets as necessary)

Originator: ____________ Signature: ____________ Date: ________

PI Committee Member:* ____________ Signature: ____________ Date: ________

* Provide notice to Potential Issue Committee Coordinator of NI initiation.

DISPOSITION (Regulatory Programs)

3. Is NI recommended to Facility Manager as a PISA? YES ☐ NO ☐

If yes, identify affected facilities and provide basis below.

If no, provide basis below and method of closure. (e.g., NCR, TSR or ORPS reference)

Regulatory Programs Manager:* ________ Signature: ____________ Date: ________

* Provide notice to Potential Issue Committee Coordinator of NI disposition.

PISA DECLARATION (Facility Manager)

4. Does the NI represent a PISA? YES ☐ NO ☐

If "Yes," refer to 11Q Procedure 1.05 for applicable actions and list the ORPS and USQE number.

If not, provide basis and obtain Facility Manager signature for closure.

Facility Manager: ____________ Signature: ____________ Date: ________

5. CLOSURE:

Regulatory Programs Manager:* ________ Signature: ____________ Date: ________

* Provide notice to Potential Issue Committee Coordinator of NI disposition.

Fig. 1. NI Processing Form.

for applicability within their Division. If the NI may be applicable, the member assigns a Division designator to the form and notifies the NI Committee Coordinator. Each Committee Member shall be satisfied that the particular NI topic has received adequate review, including trend analysis. If a Committee Member disagrees with the actions recorded on the NI Processing Form, that Committee Member will identify that disagreement to the Coordinator. The Committee Member will identify if the issue can be discussed at the next regularly scheduled meeting or if a special meeting is warranted. The Coordinator will add the item to the agenda to be discussed at the next Committee meeting. If the NI cannot be resolved within 10 calendar days from initiation, the NI Committee Coordinator informally notifies DOE-SR of the NI. This notification can be accomplished through a NI database-generated e-mail.

2. Disposition. Originator input should be utilized, as applicable, to assure accurate characterization and disposition of NI. The Regulatory Programs Section obtains a unique NI number and enters it into the tracking system. The Regulatory Programs Section notifies the NI Committee Coordinator of the NI initiation. The Regulatory Programs Section evaluates the NI in accordance with the applicable guidance provided in Procedure 1.05 of the *Facility Safety Document Manual* (WSRC 2001). A "NO" answer to the question in the "Disposition" block indicates that the NI does not require consideration as a PISA by the affected Facility Manager. A "YES" answer to the question in the "Disposition" block indicates the NI is recommended to the appropriate Facility Manager for consideration as a PISA. Once dispositioned, the Regulatory Programs Manager signs the NI Processing Form and forwards a copy of the form to the NI Committee Coordinator.

3. Declaration. A "YES" answer to the question in the "PISA Declaration" block indicates that the Facility Manager has declared a PISA in accordance with Procedure 1.05 of the *Facility Safety Document Manual* (WSRC 2001). In that event, the Occurrence Report Processing System and Unreviewed Safety Question Evaluation numbers are entered and a copy is forwarded to the New Information Committee Coordinator. If a PISA is not declared, the Facility Manager so indicates by a "NO" answer in the "PISA Declaration" block, with justification, the Regulatory Programs Manager signs the NI

Processing Form to indicate closure, and a copy is forwarded to the NI Committee Coordinator. The NI Committee Coordinator shall transmit a copy of the dispositioned NI to the Originator. If the Originator disagrees with the actions taken, the Originator may contact the NI Committee Coordinator and request that the disagreement be addressed at the next committee meeting.

The NI Committee manages the process from a process overview perspective (Fig. 2); determines when information is appropriate for entering the NI Process; coordinates the response to NI when multiple facilities are involved; tracks and follows-up on items identified; and provides technical support in the assessment of NI. The Regulatory Programs Section is ultimately responsible for processing a NI. Procedure 1.05 of the *Facility Safety Document Manual* (WSRC 2001) provides guidance in determining when a PISA exists. NI tracking is closed when the Regulatory Programs Manager signs the closure section on the NI Processing Form and the NI Committee Coordinator is notified. The Coordinator will provide periodic reports on all NI Processing Forms received. The monthly report will identify all open forms and all forms received since the last report. The report will be distributed to the Site Chief Engineer, all Division Chief Engineers, all Committee Members, and the Originators of the NI Processing Forms addressed in the report (i.e., all open forms and all new forms). NI disposition may indicate the existence of a PISA. If a PISA is declared, the Facility Manager determines whether interim facility operating restrictions are needed to place a facility or process in a safe condition. The Site Lessons Learned Program may be a source of NI originating external to a nuclear facility.

Each NI Processing Form shall be assigned a unique tracking number by the Regulatory Programs Section or the NI Committee Coordinator. NI identified as applicable to more than one Division facility will be identified by using the same sequential number (i.e., unique to the NI) with the exception that the applicable facility designator is assigned. The tracking number shall be in the following format: NI-XXX-YR-NNN, where NI is New Information Screening; XXX consists of Applicable Division/facility character alpha Division/facility designator (not limited to three characters); YR is year issued; and NNN is the sequential number assigned by the issuing organization.

NI Processing Forms constitute Quality Assurance records and must be maintained accordingly.

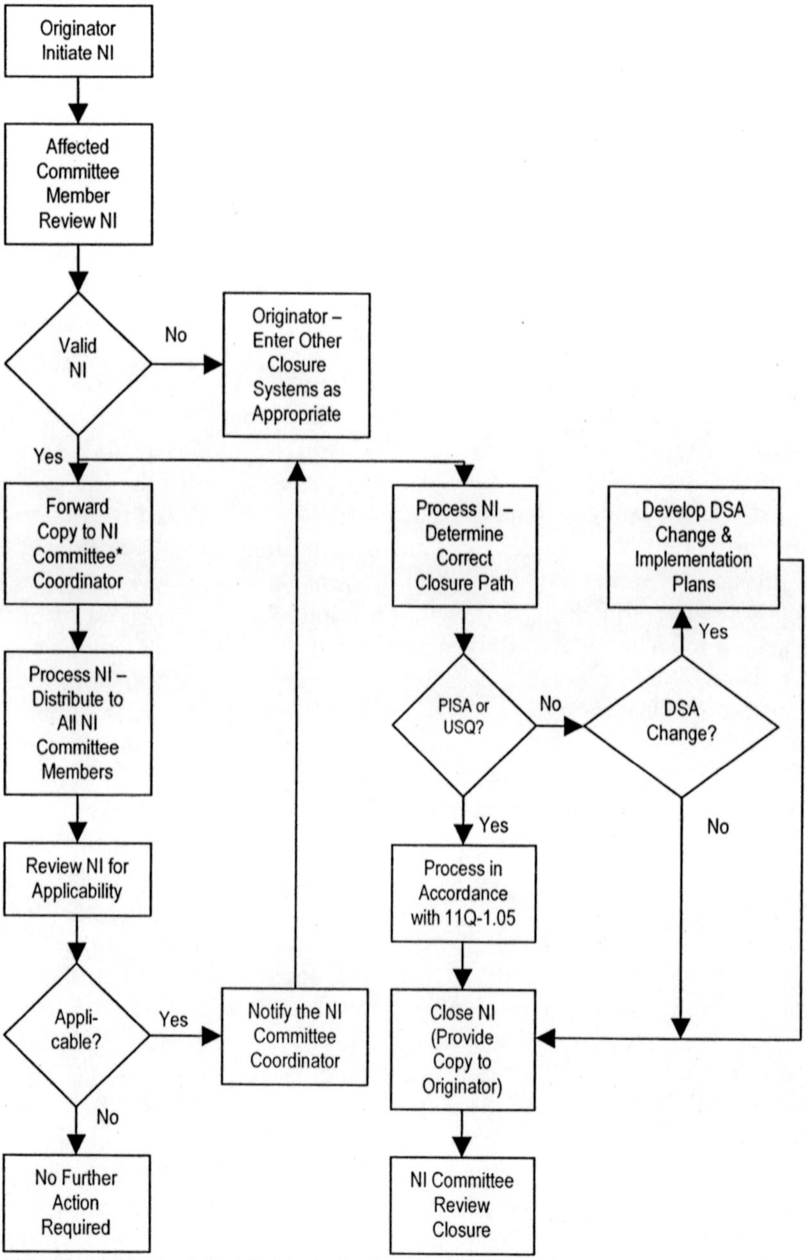

*If the NI cannot be resolved within 10 calendar days of initiation, the NI Committee Coordinator notifies DOE-SR.

Fig. 2. Processing new information to determine PISA.

FUNCTIONAL CLASSIFICATIONS

Procedure 2.25 of *Conduct of Engineering and Technical Support* (WSRC 2002c) provides the: requirements; responsibilities; and methodology for identifying Safety Class (SC) and Safety Significant (SS) functions and the structures, systems, and components (SSC) or Administrative Controls which are credited with performing these functions that reduce the risk to the Public and Workers from uncontrolled releases of radioactive and other Hazardous Materials. In addition, it provides methodology for assigning Interim, Preliminary, and Final Functional Classifications to SSCs. It applies to the classification of SSCs in existing facilities, new facilities and modifications to existing facilities. This procedure applies to all organizations that recommend, assign, or approve Functional Classifications for SSCs including:

1. Design Agencies or other Technical Agencies that recommend Functional Classifications for SSCs added or modified during the design of a modification to an existing facility or for new facilities.
2. Design Authorities that assign Functional Classifications to existing SSCs. Interim Functional Classifications and Final Functional Classifications performed to bring the facility into compliance with this procedure shall be as scheduled in the Facility Classification Plan as approved by the U.S. Department of Energy (DOE).

In addition, this procedure provides the methodology for the selection of non-SC/SS Defense-in-Depth items. Selection of an SSC as non-SC/SS Defense-in-Depth does not alter its classification. Administrative Controls are not assigned Functional Classifications. This procedure does not include performing or documenting safety analyses. Safety analyses used in implementing this procedure must be conducted in accordance with the *Facility Safety Document Manual* (WSRC 2001). The provisions of this procedure do not apply to the transportation of radiological material, offsite, onsite in-commerce, and onsite transfers as governed by the *Transportation Safety Manual* (WSRC 2002f). The provisions of this procedure, however, do apply to movement of hazardous materials or waste within the area covered by a facility safety basis or within a facility fence. The provisions of this procedure apply to WSRC and other members of the Performing Entity (i.e., Westinghouse Savannah River Company; Bechtel

Savannah River; Inc.; BWXT Savannah River Company; and BNFL Savannah River Corporation) for management and operations at Savannah River Site, and to subcontractors performing work for any member of the Performing Entity when required by contract or applicable law.

Figure 3 provides a process flow chart. All SSCs are assigned a Functional Classification of General Services (GS), unless it has been determined and documented that another Functional Classification is required. The Functional Classification of an SSC as determined by this procedure may be changed to a classification of increased precedence (e.g., SS to SC, Production Support (PS) to SS) with the documented approval of the Site Chief Engineer. The Responsible Engineer (i.e., Design Authority Engineer, Design Agency Engineer, or Technical Agency Engineer) determines whether the assignment of a higher classification than GS is necessary using the following process:

1. Determine the documentation to be used. The documentation to be used in the classification process will depend on the status of the facility or system (i.e., new, modification, or existing) and the status of the available documentation. The documents available may include, but are not limited to the Safety Analysis Report; Preliminary Hazard Analysis; Basis for Interim Operation; Safety Evaluation Report; Plant Operating Specifications; Process Hazard Reviews; Hazard Assessment Document; drawings; system manuals; test reports; Site Specifications; system operating procedure/instructions; Letters of Commitments; Justifications for Continued Operation; Hazard Analysis; Emergency Preparedness Hazard Assessments; Screening Process Hazard Reviews; Design Process Hazard Reviews; Preliminary Hazard Analyses and Process Hazard Analyses; and any interim documents that provide additional bases for facility operation. The Responsible Engineer should contact the applicable Regulatory Programs and Safety Analysis groups to determine the status of the Document Safety Analysis (DSA) and the appropriate documents to use for the Functional Classification. Functional Classification to determine SC and SS SSCs is usually an iterative process that benefits from the input of several organizations. A team approach is recommended because the Responsible Engineer should obtain input from the organizations that will use or be impacted by the

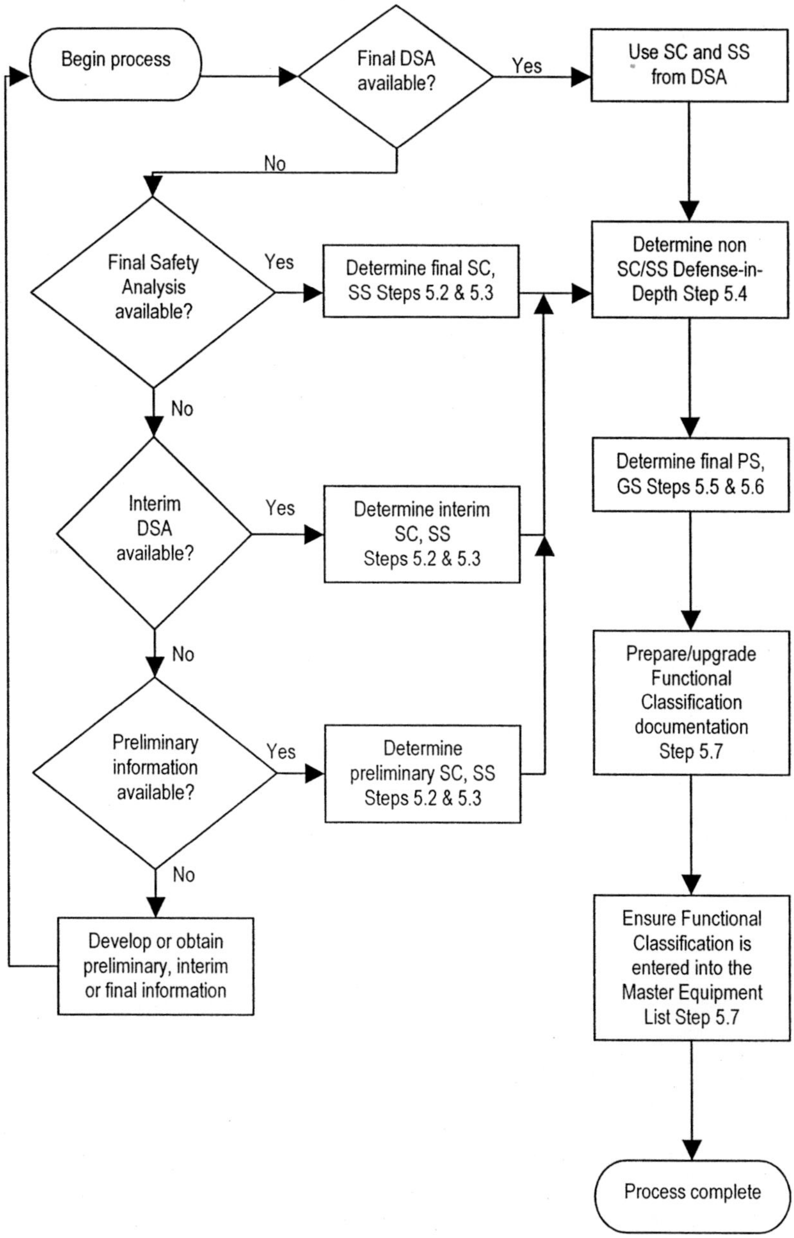

Fig. 3. Safety class criteria.

Functional Classification (i.e., Operations; Design Authority; Regulatory Programs; Safety Analysis Groups; Emergency Planning; and site fire department). The Responsible Engineer leads the selection of the proposed SSCs based on the potential exposure risk and associated operating conditions and accident scenarios as documented in the Hazard and Safety Analyses. The Responsible Engineer should contact a safety analyst when interpretations of the analyses are required.

2. Determine SSCs with a SC classification. The SC classification applies only to HC-1, HC-2, HC-3 Nuclear Facilities and their Safety Support Facilities. The Responsible Engineer reviews the documentation to determine if there is a potential for challenging SC criteria. All of the following conditions must exist to require a SC Function: 1) sufficient inventory exists that if released in total would challenge the 25 rem SC Evaluation Guideline; 2) an exposure path and mechanism for release exists; and 3) the frequency of the unmitigated event is considered credible. The Responsible Engineer identifies or determines the operating conditions and accident scenarios from the applicable hazard analysis under which the release would occur. The Responsible Engineer determines anticipated operational occurrences (e.g., startup, shutdown, specific process conditions, failure modes with credible frequencies) and accident scenarios (e.g., earthquake, tornado, process accidents) from the applicable hazard and accident analysis that could result in challenging the SC criteria. The Responsible Engineer identifies or determines those SC functions that satisfy the SC criteria. The Responsible Engineer (based on input and coordination with the accident analyst; hazard analyst; functional classification engineer; and responsible facility personnel as needed) identifies or determines those SSCs or Administrative Controls that perform the SC functions. The SC function is satisfied when the selected SC SSCs and/or Technical Safety Requirement-based Administrative Controls do not challenge the Offsite EG. For Final Functional Classifications, the Responsible Engineer provides the list of proposed SC functions and the SSCs and Administrative Controls that perform those functions to the safety analyst to be used in the mitigated safety analyses. The proposed list of SSCs and Administrative Controls are used to evaluate that the SC criteria are no longer challenged. The results of this analysis are provided to the Responsible

Engineer to either confirm the correct identification of SSCs and Administrative Controls or to revise the list. This process continues until the selection of SSCs and Administrative Controls ensures the SC criteria are no longer challenged. The SSCs selected through this process are classified SC. The Offsite Evaluation Guideline should not be viewed as a "speed limit". Offsite EGs are established for the purpose of identifying and evaluating SC controls. It is preferred that the mitigated analysis results do not challenge the guideline. However, there may be unusual cases where the cost involved in modifications to the facility required to satisfy the Evaluation Guideline can not be justified. This must be documented in the facility DSA and agreed to by DOE as indicated by the approval of the DSA.

3. Determine SSCs with a SS classification. The SS classification applies only to HC-1, HC-2, HC-3 Nuclear Facilities and their Safety Support Facilities. Worker hazards that are routinely encountered in general industry and construction, and for which national consensus codes, standards, or site programs (e.g., OSHA, transportation safety) exist to guide safe design and operation are not considered when performing SS classification of SSCs. The Responsible Engineer reviews the documentation to determine if there is a potential for challenging the SS criteria. The Responsible Engineer identifies or determines the operating conditions and accident scenarios from the applicable hazard analysis under which the material release, energy release, or inadvertent criticality accident would occur. The Responsible Engineer determines both anticipated operational occurrences (e.g., startup, shutdown, specific process conditions, failure modes with credible frequencies) and accident scenarios (e.g., earthquake, tornado, process accidents) from the applicable hazard and/or accident analysis that could result in challenging the SS criteria. The Responsible Engineer identifies or determines those SS functions that satisfy the SS criteria. The Responsible Engineer identifies or determines those SSCs or Administrative Controls that perform the SS functions. For Final Functional Classification, the Responsible Engineer provides the list of proposed SS functions and the SSCs and Administrative Controls which perform those functions, to the Safety Analysis Group to be used in the Mitigated Safety Analysis. The Safety Analysis Group uses the proposed list of SSCs and

Administrative Controls to evaluate if the SS criteria are satisfied. The results of this analysis are provided to the Responsible Engineer to either confirm the correct identification of SSCs and Administrative Controls or to revise the list. This process continues until the selection of SSCs and Administrative Controls that ensures the SS criteria are satisfied. The SSCs selected through this process are classified as SS. The SS criteria should not be viewed as a "speed limit." It is preferred that the mitigated analysis results do not challenge the criteria. However, there may be unusual cases where the cost involved in modifications to the facility required to satisfy the SS criteria can not be justified. This must be documented in the facility DSA and agreed to by DOE as indicated by the approval of the DSA.

4. Determine SSCs with non-SC/SS Defense-in-Depth functions. All the SSCs and Administrative Controls that can act to protect either the public or workers from the release of a hazard by protecting against uncontrolled releases of Hazardous Material are part of a broad picture of the facility's Defense-in-Depth. In the Functional Classification process, some of these overall Defense-in-Depth items are selected as: SC to protect the Public; SS to protect the Public; SS to protect the Workers; SS to provide Defense-in-Depth to the items selected above; SS to prevent, monitor, or detect a nuclear criticality; or Non SC/SS Defense-in-Depth to ensure the facility satisfies the Non SC/SS Defense-in-Depth goals. These items are not classified as either SC or SS. They are classified as either PS or GS. The Responsible Engineer reviews the accident scenarios to determine all accidents where the unmitigated consequences challenged the offsite criteria. For these accidents, the Responsible Engineer evaluates the mitigated consequences to the public crediting SC controls and qualitatively crediting SS controls. If these mitigated consequences exceed the goals, additional SSCs and/or Administrative Controls must be identified to move the consequences closer to these goals. These additional SSCs and Administrative Controls will be selected from the mitigators and preventers that were not previously selected to provide a SC or SS function. These will be referred to as non-SC/SS Defense-in-Depth. A qualitative evaluation of risk, considering all SC, SS, and non-SC/SS Defense-in-Depth controls, will be performed to support the selections. The detailed evaluation will be documented

in an engineering calculation. The conclusions of the evaluation will be documented in a Technical Report. The Technical Report will not be approved by DOE and will, therefore, not be part of the DSA. The list of non-SC/SS Defense-in-Depth SSCs and Administrative Controls will be documented in a DSA, along with reference to the above-identified Technical Report.

5. Determine SSCs with a PS classification. The PS classification applies to both nuclear and non-nuclear facilities.

The GS classification applies to both nuclear and non-nuclear facilities. At a minimum, all Savannah River SSCs are classified as GS. SSCs are reclassified to one of the other classifications only as required to satisfy one of the criteria for those classifications. All SSCs must fulfill the GS functions required to ensure the health and safety of the worker as identified in DOE orders, selected National Consensus Codes and Standards, and Site Standards. During the Functional Classification process, SSCs can be classified with varying levels of detail. For instance, when the classification process is initially performed for a facility and a system is classified as SC or SS, the process may stop at that level and all components will be assigned the classification of the system. The process may continue to classify the supporting systems and components separately. The process of classifying the supporting systems and components may be performed over a longer period with each item individually classified only as needed (e.g., a component in a system may not be initially classified individually, but will be classified if there is a need to "touch" the item in the maintenance or replacement arena). This process also applies to components and their sub-components. The classification information is approved by the authoring organization manager and the Design Authority.

Each facility will generally have system level Functional Classifications documented in a technical report. This may be a stand-alone document or the comparable information may be contained in another document(s) such as a Hazard Analysis (HA) or DSA. This information will correlate accident/events from the HA/DSA to the specific SSCs (at the system level) credited for the accident/events. It will provide a clear picture of how the facility Functional Classification meets the requirements of this procedure, including information on logic and technical justification for the results.

Component-level classification documentation must contain the information described in this section. For SSCs added or modified by a plant modification or as part of a new facility, the Design Agency or other Technical Agency documents the proposed Functional Classification. The proposed Functional Classifications are reviewed and approved by the Design Authority during the normal review and approval process of the design output documents. As a minimum, the functional classification documentation shall include the following:

1. SSC name
2. SSC number
3. Reference to the documentation that justifies the functional classification of SC, SS, or PS
4. Functional classification. Each SSC may provide more than one SC, SS, PS, or GS function. The SSC is assigned a functional classification according to the order of precedence (i.e., SC, SS, PS, GS).
5. For SC, SS, and PS classified SSCs, identification of: each SC, SS, and PS function required to be performed (a single SSC may have multiple functions); and conditions under which each SC, SS, and PS function is required (operating conditions and accident scenarios).
6. For SC and SS SSCs, the status of the classification (i.e., preliminary, interim, or final)
7. Design Authority approval

For Preliminary Status Functional Classification of PS and GS SSCs, the following information is not required:

1. Whether it is DSA related or not
2. The Component Location Identifier
3. Integrated Data Processing Number

A revision to an FCD that has been submitted to Document Control is processed by direct revision and approved in accordance with the approval process for the original FCD. Prior to submittal to Document Control, minor administrative changes to the completed FCD may be made by drawing a single line through the error, recording the correction information, initialing and dating the correction. The FCD number must reflect the current revision. The revised FCD or equivalent is forwarded to Document Control. If the SSC/part Functional Classification has been changed, the Responsible Engineer ensures that the revised Functional Classification is entered into the Master Equipment List. Functional Classifications are documented in QA Records.

NUCLEAR FACILITY UNREVIEWED SAFETY QUESTIONS

Procedure 1.05 of the *Facility Safety Document Manual* (WSRC 2001) is to be performed for any Proposed Activity (PA). It defines responsibilities to implement the requirements of 10 CFR 830.203 (DOE 2001a). The primary role of the Unreviewed Safety Questions (USQ) Process is to preserve the Safety Basis (SB) for each Nuclear Facility while allowing for operational flexibility. The USQ Process determines if a PA involves a USQ. The procedure establishes a two-part process that includes:

1. USQ Screening (USQS) to identify those PAs that require a USQ Evaluation (USQE)
2. USQ Evaluation

This procedure also addresses initiation of a change to the SB, if required. In addition to a PA, the procedure describes a second type of USQ dealing with the discovery of an inadequacy in the Safety Analysis for a facility. The provisions of this procedure apply to Westinghouse Savannah River Company (WSRC) and other members of the Performing Entity (i.e., WSRC; Bechtel Savannah River, Inc.; BWX Technologies, Inc. Savannah River Company; and British Nuclear Fuels Limited Savannah River Corporation) for management and operations at Savannah River Site (SRS), and to subcontractors performing work for any member of the Performing Entity when required by contract or applicable law. The requirements herein are applicable to the Department of Energy owned HC-1, HC-2, HC-3 Nuclear Facilities at SRS operated by WSRC. The USQ Process is applicable when a Documented Safety Analysis (DSA) has been established by WSRC for a Nuclear Facility. 10 CFR 830.203 (DOE 2001a) uses "probability of an accident" as a criterion for USQ determination. Probability of an event depends on the life of the facility. Since the facility life is unknown, at SRS the frequency or the number of events per year is used in the USQ process.

An Unreviewed Safety Question is a situation that has not been previously reviewed or approved by DOE, and meets one or more of the following conditions as in 10 CFR 830, Subpart B (DOE 2001a):

1. The frequency of occurrence of an accident that was previously evaluated in the Documented Safety Analysis may be increased
2. The consequences of an accident that was previously evaluated in the Documented Safety Analysis may be increased
3. The frequency of occurrence of a malfunction of Important To Safety Equipment that was previously evaluated in the Documented Safety Analysis may be increased
4. The consequences of a malfunction of Important To Safety Equipment that was previously evaluated in the Documented Safety Analysis may be increased or
5. The possibility for an accident of a different type than any evaluated previously in the Documented Safety Analysis may be increased
6. The possibility of a malfunction of Important To Safety Equipment of a different type than any previously evaluated in the Documented Safety Analysis may be increased
7. The Margin of Safety as defined in the basis for any Technical Safety Requirements or other technical analysis that define the safety limits could be reduced
8. The Documented Safety Analysis may not be bounding or may be otherwise inadequate.

The Facilities Operations Safety Committee (FOSC) or equivalent committees such as the Reactor Operation Safety Committee is responsible for reviewing and approving all USQEs. The Facility Manager (for a Hazard category 1, 2 or 3 Nuclear Facility) is responsible for:

1. Ensuring that all PAs in the facility are reviewed for USQ
2. Designating as USQ Screening Originator (SO); USQ Screening Reviewer (SR); USQ Evaluation Originator (EO); and USQ Qualified Reviewer (QR) individuals who meet the qualification and training requirements
3. Providing resolution to FOSC comments
4. Resolving the disagreements between the SO, SR, EO, and the QR
5. Ensuring that existing and new procedures that are sources of PAs are integrated into the USQ Process thereby ensuring that PAs are properly evaluated by the USQ Process
6. Notifying DOE about discoveries of a Potential Inadequacy in the Safety Analysis

7. Ensuring that at no time during the course of implementing a modification is the facility placed in a configuration that was not evaluated by the USQ Process
8. Submitting the Safety Analyses and USQ documentation packages to DOE as appropriate

The Environment, Safety, Health, and Quality Assurance Division will evaluate the USQ Process from an oversight capacity.

The QR is an individual assigned by facility management who meets the qualification and training requirements. The review performed by the QR is considered a Technical Review of the USQS or USQE. The QR will not have directly participated in the preparation of the Proposed Activity or the USQE to be reviewed and is responsible for:

1. Being cognizant of the relevant SB and approved PAs awaiting incorporation into the SB prior to initiating the review
2. Reviewing the USQS or the USQE as requested
3. Verifying that the documentation referenced in the USQS or USQE is appropriate
4. Verifying that the forms were filled out accurately and completely
5. Independently agreeing or disagreeing with the findings made by the EO
6. Disqualifying himself/herself as QR if the review would not be independent

The Regulatory Programs Section, or equivalent organization as assigned by the applicable Division management, is responsible for:

1. Tracking approved SB changes that are pending incorporation into the SB
2. Developing and maintaining a list of Categorical Exclusions and forwarding copies of the USQEs for Categorical Exclusions to DOE-SRS for information
3. Reporting to DOE all situations for which a USQE was required

Any reference to Regulatory Program in this procedure applies to the equivalent organization.

The SO is an individual who meets the qualification and training requirements, was assigned technical responsibility by facility management, and who is responsible for:

1. Being knowledgeable of the relevant SB and approved PAs awaiting incorporation into the SB prior to initiating a USQ Screening
2. Performing the USQS
3. Initiating changes to SB documents that are identified as a result of the USQ process

The SR is an individual who meets the qualification and training requirements, was assigned technical responsibility by facility management, and who is responsible for:

1. Being knowledgeable of the relevant SB and approved PAs awaiting incorporation into the SB prior to initiating a USQ Screening
2. Performing the USQS Reviews; the SO and SR are not qualified to prepare or review USQEs

The EO is an individual who meets the qualification and training requirements of this procedure, was assigned technical responsibility by facility management to perform USQEs, and who is responsible for:

1. Being knowledgeable of the relevant SB and approved PAs awaiting incorporation into the SB prior to initiating a USQE
2. Performing the USQE
3. Initiating changes to SB documents that are identified as a result of the USQ Process

EO and QR are also qualified to perform, if needed, the USQS or review the USQS prepared by other individuals.

The Safety Analyst is responsible for completing the Safety Analysis and/or reviewing it for technical adequacy when requested (the Safety Analysis shall not be completed and reviewed by the same individual).

Events reported in Documented Safety Analysis are divided into four frequency categories, as follows:

Frequency Category	Frequency Range (per year)
Beyond Extremely Unlikely	frequency $< 10^{-6}$
Extremely Unlikely	$10^{-6} <$ frequency $< 10^{-4}$
Unlikely	$10^{-4} <$ frequency $< 10^{-2}$
Anticipated	frequency $> 10^{-2}$

Figure 4 provides a general overview of the USQ Process. The USQ Form (Fig. 5) is used to document the review of a PA for USQ. A completed USQS constitutes the USQ documentation if no USQE is required.

The USQ Process shall be initiated for, but not limited to, the following:

1. Permanent or temporary hardware changes to a facility
2. Permanent or temporary procedure changes
3. New operations, tests or experiments
4. Potential Inadequacy in Safety Analysis and
5. Discovery of a discrepancy between the facility physical configuration and that assumed in the DSA if it is desired to retain the as-built configuration

Eliminating or modifying a DSA identified Non-Safety Class (SC)/Safety Significant (SS) Defense-in-Depth (DID) control represents a change in commitment made in the DSA; therefore, the DOE needs to be made aware of the change. Eliminating or modifying the Non-SC/SS Defense-in-Depth controls will not in itself result in a USQ. "Modifying" as used here, is limited to a change that would affect the performance of its DSA defined DID function. When the Design Authority reviews the proposed change, a USQS or USQE shall be performed as necessary. If the USQ process does not result in a USQ, then a copy of the USQ documentation, clearly identifying the impact to the DID control shall be forwarded to DOE for information.

The SO, SR, EO, and QR shall be cognizant of the relevant SB associated with the activity. In evaluating the PA, the SO, and EO shall treat any intermediate configurations of the processing facility area as a part of the activity. There are special cases where a change to the facility is not involved and a USQS and/or a USQE is still required. Examples of these cases are:

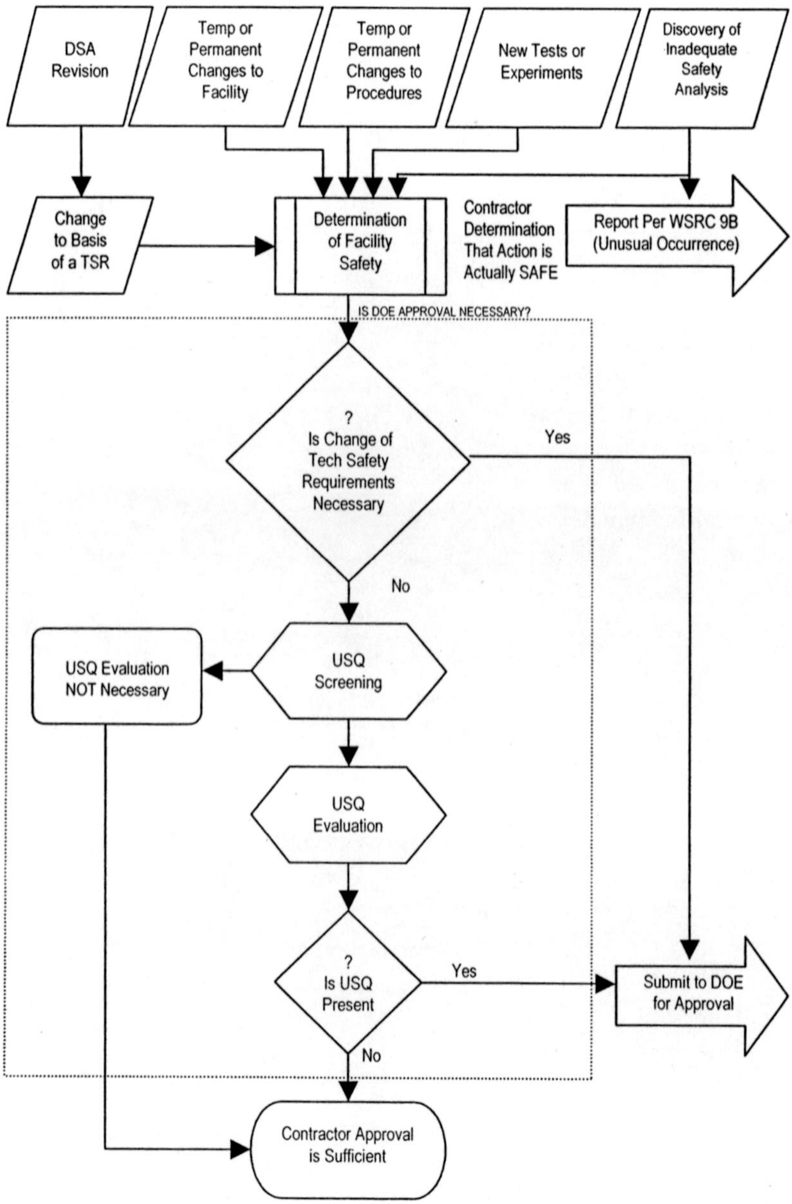

Fig. 4. Overview of USQ process.

UNREVIEWED SAFETY QUESTION PROCESS
USQ SCREENING - PART A

USQ No. ______________ Page 1 of 3

Title: ________________

Description of Proposed Activity* (or Discovery):

____

____

* Includes intermediate configurations and impacts on other facilities which might result from the proposed activity

1 SCREENING ORIGINATOR

Is the Proposed Activity a change to TSRs/OSRs? ❒ NO ❒ YES

Justification and References:

____

____

If YES, prior DOE approval through the TSR/OSR change process is required, no further USQ screening or evaluation is required. GO TO Block 3 and 4 and complete. IF NO, continue with screening.

2

Does the Proposed Activity involve:

a. Change to the facility as described in the Documented Safety Analysis? ❒ NO ❒ YES

b. Change to the procedures as described in the Documented Safety Analysis? ❒ NO ❒ YES

c. Test or experiment not described in the Documented Safety Analysis? ❒ NO ❒ YES

d. Analytical errors, omissions or deficiencies in the Documented Safety Analysis? ❒ NO ❒ YES

If either question a, b, c, or d is answered YES, justification below is not required, complete Blocks 3 and complete a USQ Evaluation (Block 5)

Supporting Information

References (SB documents considered):

____

____

Justification (Required if response to ALL questions above are

____

____

____

3 SCREENING ORIGINATOR

a. Is a USQE required? (If Yes submit to EO for USQE) ❒ NO ❒ YES

b. If a USQE is not required, does the PA require a change to the SB in accordance with 11Q? ❒ N/A ❒ NO ❒ YES
(If YES, forward a copy of the USQS to Regulatory Programs)

c. Does this PA eliminate or modify a DSA identified Non-SC/SS Defense-In-Depth Control? ❒ NO ❒ YES
(If YES, forward a copy of the USQS to Regulatory Programs for transmittal to DOE)

Comments:

Signature /Print Name	Date	Location	Department	Phone No.

4 SCREENING REVIEWER

Is a USQE required? ❒ NO ❒ YES

Comments:

____

____

Returned to SO for: ❒ Initiation of USQE Process ❒ Implementation of PA ❒ Initiation of TSR/OSR Change

Signature /Print Name	Date	Location	Department	Phone No.

Fig. 5. Template for unreviewed safety question process.

UNREVIEWED SAFETY QUESTION PROCESS
USQ EVALUATION - PART B

USQ No. ______________ Page 2 of 3

5 USQ EVALUATION ORIGINATOR

a. Could the Proposed Activity increase the frequency of occurrence of an accident previously evaluated in the Facility Documented Safety Analysis? ❒ NO ❒ YES

Justification/References:

b. Could the Proposed Activity increase the consequences for an accident previously evaluated in the Facility Documented Safety Analysis? ❒ NO ❒ YES

Justification/References:

c. Could the Proposed Activity increase the frequency of occurrence of a malfunction of equipment important to Safety previously evaluated in the documented Safety Analysis? ❒ NO ❒ YES

Justification/References:

d. Could the Proposed Activity increase the consequences of a malfunction of equipment Important to Safety previously evaluated in the Facility Documented Safety Analysis? ❒ NO ❒ YES

Justification/References:

e. Could the Proposed Activity create the possibility for an accident of a different type previously evaluated in the Facility Documented Safety Analysis? ❒ NO ❒ YES

Justification/References:

f. Could the Proposed Activity create the possibility of a malfunction of equipment Important to Safety of a different type than previously evaluated in the Facility Documented Safety Analysis? ❒ NO ❒ YES

Justification/References:

g. Could the Proposed Activity decrease the Margin of Safety as defined in the basis for any Technical Safety requirement or other technical analyses that define the Safety Limits in the Facility Safety Basis (e.g. TSRs/OSRs, Tech, Standards, JCOs)? ❒ NO ❒ YES

Justification/References:

6 USQ EVALUATION ORIGINATOR

Is a USQ involved? ❒ NO ❒ YES

Signature	/Print Name	Date	Location	Department	Phone No.

7 QUALIFIED REVIEWER

Is a USQ involved? ❒ NO ❒ YES

Comments:

Returned to EO for: ❒ Presentation to FOSC ❒ For resolution of comments

Signature	/Print Name	Date	Location	Department	Phone No.

Fig. 5. (cont'd).

UNREVIEWED SAFETY QUESTION PROCESS
USQ EVALUATION - PART B (continued)

USQ No. ______________ Page 3 of 3

8 FOSC CHAIRPERSON

Is a USQ involved? ❒ NO ❒ YES
(If No, return to EO for implementation of the PA)

Does the PA require a change to the SB in accordance with 11Q? ❒ NO ❒ YES
(If Yes, forward a copy of the USQE to Regulatory Programs)

Is the USQE for the approval of Categorical Exclusion? ❒ NO ❒ YES
(If Yes, forward a copy of the USQE to Regulatory Programs)

Comments:

____

____

If a USQ is involved indicate the appropriate follow-up action by checking one of the following boxes and return to EO

❒ Cancel PA ❒ Modify PA to attempt to eliminate USQ ❒ Initiate request for DOE approval of the PA

Signature	/Print Name	Date	Location	Department	Phone No.

Fig. 5. (cont'd).

1. When the as-built condition of the facility is not as described in the SB and it is desired to retain the as-built configuration. This constitutes a change to the facility as described in the SB and, therefore requires a USQE, even though no physical change took place in the facility.
2. When the identification of a situation indicates a Potential Inadequacy in the Safety Analysis.

The USQ process alone shall not be used to delete information from the SB. Information may be deleted from the SB via the USQ process if this is the direct result of a change to the facility or procedures evaluated in accordance with the USQ process. While completing the USQS/USQE, the SO, SR, EO, or QR may identify cases in which SB documents need to be revised. A revision may be needed even though a USQ does not result from the USQE. To identify these potential revisions, the following activities must be completed:

1. In preparing the USQS/USQE, the appropriate sections of the SB documents must be reviewed and referenced. Changes resulting from a PA or discrepancies between the facility, process, or equipment descriptions in the DSA documents and actual conditions must be identified. This does not imply that a USQ exists. It simply means

that the safety documentation does not describe adequately the existing or proposed condition.

2. If a discrepancy or potential change is identified in the SB documents, the SO/EO shall initiate a SB document change in accordance with appropriate Procedures and forward the change for review to the Regulatory Programs Section (or other groups as designated in the Division/Department procedure).
3. All USQEs will be forwarded to the Regulatory Programs Section.

In principle, new analyses and studies (e.g., safety, seismic, criticality) may be performed at any time during a facility's life-cycle. It is important that the results of these new analyses be properly reviewed for impact on the facility's SB:

1. Changes made external to a facility (e.g., changes to site infrastructure such as road closures, controlled burns, changes to fire protection systems, etc.) shall be reviewed.
2. New analyses shall be made available to all parties having an interest in the safety of the affected facility or any facility where interaction occurs. Included, for example, are the affected facility(s) manager, facility technology and engineering groups, regulatory programs sections, and the Safety Analyst.
3. The appropriate line manager is responsible for: ensuring that the analyses are reviewed for impacts on the facility's SB documents; and properly processing the NI.
4. The Safety Analyst is responsible for reviewing the analyses, studies and, any USQEs for which a USQ is indicated and is responsible for initiating any required changes into the affected SB documents.

New and revised procedures, even if intended to be only temporary in duration, must have a USQS completed, as a minimum. Whenever a PA is found to involve a USQ, that PA cannot be implemented until DOE approval has been obtained. DOE approval shall be obtained via formal request consisting of:

1. USQ documentation
2. Information from the PA documentation as necessary to support the request, including the associated Safety Analysis
3. Formal transmittal letter from the respective Divisional Management level

A PA found to involve a USQ shall be reviewed by a FOSC and Division Manager or designee before transmittal of the DOE Approval Request. After the USQ documentation has been approved, any change to the PA requires the USQS and/or USQE to be revised, as necessary. The original USQ tracking number should be retained and the revised USQS and/or USQE processed.

The traditional application of the USQ Process is associated with a proposed (i.e., future) activity. However, 10 CFR 830.203 (DOE 2001a) adds a second type of USQ, "Discovery of Inadequacy in the Safety Analysis," that for brevity, herein, is called a "Discovery USQ." This type of USQ involves the discovery of New Information that indicates there may be an inadequacy in the Safety Analysis supporting the current SB for a facility. The inadequacy must be significant in terms of an increase in the risks of facility operation beyond that described in the current SB. The USQE process is used to measure significance. It is important to distinguish between the concept of a Discovery USQ and "operating events". Not every event that results in exceeding the boundaries of a Safety Analysis is a Discovery USQ. For example, the occurrence of an operational event that results in exceeding the limits specified in the SB would NOT be a Discovery USQ, if the event was caused by failure to follow operating procedures. A Discovery USQ deals specifically with the question of validity of the Safety Analysis that supports the DSA. Of course, certain types of operating events may add New Information that brings into question the validity of risks presented in the Safety Analysis. The discovery of New Information that the current Safety Analysis is potentially inadequate is expected from several general sources: discovery of errors, inaccuracy, or omission in the current Safety Analysis, including discovery of a potential failure mechanism that had not been previously known or considered; a vendor notification that certain equipment experiences failure under certain conditions; technological advances revealing that important assumptions in the Safety Analysis are invalid; the discovery of a discrepant "as-found" state (e.g., actual facility physical configuration and the configuration assumed in the DSA do not agree); and the occurrence of an event. Certain New Information received as part of a Safety Analysis upgrade is excluded from being considered as a Discovery USQ. An example is new Safety Analysis results stemming from the application of new DOE requirements. For example, DOE may require the addition of Beyond Design Basis Accidents in the DSA. These new results would not invalidate the current Safety Analysis since the new requirement is beyond the current DSA. New Information identified as a result of an operational event may in some cases be excluded. An example is where the event did not result in exceeding a Technical Safety Requirement (TSR) Safety Limit. Another example is where failure to comply with specific procedures and operational conditions create an event beyond the safety envelope. In this situation, the current analysis remains valid since the cause of exceeding the safety envelope is tied to a programmatic or procedural breakdown, not analytic inadequacies. A potential source for exclusion is when New Information involves primarily the performance capability of TSR related facility hardware. If the performance degradation of the facility hardware could cause noncompliance with a Limiting Condition of Operation in a TSR, the appropriate actions per the facility specific TSRs and procedural actions shall be taken. In this case, a USQ review would not be necessary because actions are taken in accordance with the SB to assure the continuing validity of the current DSA. Not every confirmed error or deficiency with a Safety Analysis will be of enough significance to warrant the declaration of a Discovery USQ. A deficiency must be tested for significance using the USQE process to determine if a USQ exists. For example, a discovered analytical error that the consequences of radioactive release in an accident scenario reported in the DSA should have been greater by a factor of two, may (or may not) become a Discovery USQ. The determination is made by applying the greater consequence value to the USQE process, which is based on the seven USQ questions and related numerical criteria. In general, a useful approach is to "look back" and apply the New Information as if it were a proposed activity. For example, if New Information reveals that the performance of equipment Important to Safety will not meet design requirements, the new (lower) level of performance is applied in a safety re-analysis. These new results are then applied to the USQE to test significance. If a USQ results, the situation will be reported as a (Discovery) USQ. Note from this example that decisions related to future plant operation will have no impact on this "Discovery USQ" question. That is, if in this case a decision is made not to operate the facility in the future until new equipment that meets the stated performance requirements is installed, the Discovery USQ declaration would be unaffected.

Depending on the complexity of a situation, a considerable time-lapse can exist between the initial receipt of New Information and the final determination of whether a Discovery USQ exists. The intent of 10 CFR 830.203 (DOE 2001a) is to avoid excessive time between the receipt of New Information and notification to DOE. On the other hand, notifying DOE and taking actions based on unconfirmed information is not desirable. In view of the above, the term "Potential Inadequacy in Safety Analysis" is designated as an intermediate point in the total information evaluation process that precedes the declaration of a Discovery USQ, but for which sufficient "potential" exists for a USQ to warrant taking certain precautionary actions. Before a Potential Inadequacy in the Safety Analysis is declared, a reasonable time period is permitted to confirm the reliability of New Information and assess the potential for a Discovery USQ. It is recognized that identifying this point in the process can not be precisely defined; rather, it is based on the judgment of facility management, considering all of the circumstances and factors involved.

In accordance with 10 CFR 830.203 (DOE 2001a) and upon declaration of a "Potential Inadequacy in Safety Analysis", the following actions are required:

1. Take appropriate action to place or maintain the facility in a safe condition. Determine if interim facility operating restrictions to place the facility/process in a safe condition are necessary. Upon identification, it is incumbent on the facility management to communicate its concerns and determination of whether interim actions are necessary to DOE.
2. Communicate the situation to DOE. A Justification for Continued Operation (JCO) may be used as a means to obtain DOE approval of continued operations of a facility during the evaluation period.
3. Perform a USQE and promptly notify DOE of the results.
4. Submit the evaluation of the safety of the situation to DOE prior to removing any operational restriction initiated to place the facility in a safe condition.

In following the declaration of a Discovery USQ, the following actions are required:

1. The existence of a USQ must be formally reported to DOE. If necessary, upgrade the occurrence report to an Unusual Occurrence.
2. The completed USQE must be sent to DOE.

DOE approval is required prior to removing any operating restrictions needed to place the facility/process in a safe condition. A JCO should be used as a means to obtain DOE approval of temporary operations of a facility beyond the current SB.

The USQ Screening (USQS) requires that a description of the PA or discovery be entered on the USQS sheet of the USQ Form. This description shall contain adequate detail to support the conclusions reached in each of the following sections of the form. Reference documents and revision numbers, including technical reports, engineering drawings, procedures, and notebooks shall be listed to support the PA description as applicable. Additional sheet(s) shall be used and attached if the space provided is not adequate. Conclusions for each of the USQS questions are recorded on the USQS form by the SO. The SO shall consider all WSRC approved changes awaiting incorporation into the SB that are relevant to the PA. This may be accomplished by contacting the Regulatory Programs Section for WSRC-approved SB changes pending incorporation in the SB. Additionally, it is equally important that when new or modified PAs may have an impact on another facility's SB, a review should be completed when that interface is not defined by existing documents. This may be accomplished by submitting the supporting documentation to the NI Committee. The EO and QR are also qualified to perform, if needed, the USQS or review the USQS prepared by other individuals.

A USQE involves a series of steps and approval actions. The final result of the USQ process is a determination of whether a USQ exists. When a USQE is required, the USQE sheet of the USQ Form is filled out by the EO to document the USQE. A USQE may be performed without a USQS, even if a USQS indicates the USQE is not required or it is apparent that a USQS would require completion of a USQE. The EO checks the appropriate box and then supports the conclusion in the space provided, including references. A continuation sheet(s) must be attached if the space provided is not adequate. If the EO answered each question in section 5 "NO", this indicates a USQ does not exist for the PA. A "YES" answer to any question on the USQE indicates that a USQ may exist for the PA. This should be so noted on the form, signed, and forwarded to the QR. If the QR concurs with the EO's USQE, the QR indicates concurrence in block 7, and forwards the USQE to the EO for review and approval by the FOSC. Whenever the QR disagrees with the finding(s) of the EO, the QR returns the USQE with comments to the EO for resolution. If these comments cannot be resolved, the EO so indicates on the USQE and forwards it to Facility Manager for review and disposition.

When the applicable FOSC receives the USQE, the committee reviews the documentation and actions are taken as follows:

1. If the FOSC has comments or questions, it identifies them and returns the package to the EO for resolution.
2. If the PA does not involve a USQ, the FOSC so indicates and forwards the form back to the EO for implementation of the PA.
3. If a USQ is involved for the PA, the FOSC forwards a request to the EO for either cancellation; modification such that it no longer would involve a USQ; or preparation of a DOE Approval Request, including the performance of any analysis needed to support such a request.
4. If a DOE Approval Request is to be prepared, the EO initiates the preparation of any requests for assistance and follows the performance of any analysis to completion.
5. The EO presents the PA and USQE to the Division Manager/ designee for transmittal to DOE.
6. An "official USQ" exists when all necessary WSRC reviews and approvals have been obtained.

For PAs that would not involve a USQ, the PA can be implemented without prior DOE approval. A determination that a PA involves a USQ does not mean the PA cannot be performed in a safe manner. It means that the safety implications of the PA have not been reviewed previously by DOE. Thus, DOE approval is required prior to implementation.

Conversely, a determination that a PA does not involve a USQ does not mean there are no safety implications associated with the PA. It means that any accidents or malfunctions associated with the PA are bounded by those previously reviewed by DOE, and thus prior DOE approval is not required. It is important that the USQE contain sufficient detail so an independent reviewer can reach the same conclusion.

Since the DSA updates must be submitted to DOE annually, at any given time the version of the DSA last submitted to DOE may not reflect the most recently implemented PAs. For purposes of conducting USQEs, the EO and QR shall consider all changes awaiting incorporation into DSA document updates that are relevant to the PA. When assessing the effects the PA may have on frequencies, consequences, or margins of safety reported in the DSA, the accumulative effects of all WSRC approved changes awaiting incorporation into the DSA document updates (except for OSRs/TSRs) shall be compared to the frequencies, the consequences, or the margins of safety (respectively) of individual accidents reported in the current DOE approved DSA documents.

NUCLEAR CRITICALITY SAFETY

The purpose of the *Westinghouse Savannah River Company Nuclear Criticality Safety Manual* (WSRC 2002b) is to: define and establish a company-wide Westinghouse Savannah River Company (WSRC) nuclear criticality safety program consistent with applicable U.S. Department of Energy (DOE) requirements, industry standards, company safety policy, and accepted safety practice; provide interpretation and guidance for the uniform implementation of these requirements and standards at the Savannah River Site (SRS); and serve as the basis for criticality safety implementing procedures and manuals at the Division or lower levels of the organization. The Manual addresses the nuclear criticality safety requirements for WSRC facilities that contain significant quantities of fissionable material. The Manual establishes WSRC criticality safety requirements and provides interpretation and guidance, where necessary, for consistent implementation of criticality safety requirements within WSRC. The Manual is not intended to contain an exhaustive compilation of criticality safety requirements and good practices for every situation and every facility operated by WSRC. Rather, the Manual defines the overall site program and serves as the basis for the development of facility procedures and division/area manuals that contain criticality safety requirements and good practices, tailored as appropriate, throughout WSRC. WSRC design, operating, technical support, and oversight organizational units are responsible for implementing the requirements of this Manual, as applicable.

As noted in DOE-STD-1027 (DOE 1997), Hazard Category 3 facilities, by definition, do not contain sufficient fissile material to present a criticality hazard and are not subject to criticality safety controls, except that inventory limits specified in the facility Technical Safety Requirements (TSRs) may be necessary to control and limit the quantity of fissile material. The primary nuclides of concern which are covered by the WSRC criticality safety program are ^{235}U and ^{239}Pu. Other nuclides which require criticality safety consideration or control are: ^{233}U, ^{237}Np, ^{238}Pu, ^{240}Pu, ^{241}Pu, ^{242}Pu; various Am nuclides; and various Cm nuclides. The nuclear criticality program is not required to address sabotage or other malevolent acts. DOE Order 5480.23 (DOE 1994) specifically states that such acts are not required to be addressed in safety analysis reports. Thus, sabotage is not addressed as part of typical facility nuclear safety programs. Rather, sabotage is dealt with as part of DOE safeguards and physical protection requirements.

The Manual was developed based on existing DOE Orders, DOE Standards, industry standards commonly used throughout the DOE complex, and applicable WSRC policies and procedures. Information which forms the basis for the Manual includes: DOE (2000b); DOE (2001b); DOE (1998); DOE (2000a); ANS (1981; 1983; 1984; 1991; 1995; 1996a,b; 1997a,b; 1998a,b); WSRC (2001;2002a), Westinghouse Safety Management Solutions (WSMS) (2001).

Applicability of the requirements of this Manual spans design, construction, operation, maintenance, and decommissioning of covered facilities. The requirements of this Manual which implement current DOE requirements apply regardless of the initial facility startup date unless documented exemptions are obtained for both the DOE requirements and this Manual. Subsequent changes, additions, or deletions to the Manual will apply to operations, systems, and facilities designed or significantly modified after the date of such Manual changes unless specifically designated as a retrofit activity by DOE or this Manual. It is recognized that the design and as-built configuration of some existing facilities do not meet all the requirements of this Manual or the DOE Orders because they were built prior to the development of certain DOE Orders. Where this is the case, a Basis for Interim Operation or Justification of Continued Operation shall be on file or shall be prepared and deviations from the requirements analyzed in a facility SAR revision or addendum. Where practicable, currently existing operations, systems, and facilities which do not satisfy the Manual requirements should be upgraded. The practicability of such upgrades should consider the cost versus the benefits. Assuming cost effectiveness, upgrades of a currently existing operation, system, or facility should be approved if they improve criticality safety even if they may not result in complete compliance with this Manual. Requirements of this manual apply to all WSRC employees and WSMS employees which support WSRC. All non-WSRC employees who work in or visit areas affected by this manual shall either be escorted by qualified personnel or appropriately trained for compliance with minimum area-specific criticality safety requirements such as knowledge of the sound of the Nuclear Incident

Monitor (NIM) alarm, the nuclear incident evacuation routes, and the location(s) of associated rally points.

Under the SRC Site Chief Engineer's direction, the Criticality and Radiation Transport Section of the Westinghouse Safety Management Solutions affiliate (WSMS/CRT) is the cognizant organization responsible for the day-to-day interpretation of this Manual, consistent with the WSMS/CRT criticality safety programmatic responsibilities. If agreement between WSMS/CRT and another division cannot be achieved, the matter may be appealed to the Nuclear Criticality Safety Review Committee (NCSRC) and ultimately the Technical and Engineering Management Integration Council (TEMIC) for resolution. Clarification may be requested from DOE (for issues involving DOE Orders and Standards). The NCSRC is responsible for approval of revisions to this Manual. WSRC is in the process of a transition from Operational Safety Requirements (OSRs) to TSRs. However, some facilities will retain OSRs and will not prepare TSRs. Division, Department, or any other lower-level criticality safety manuals and procedures shall be consistent with this Manual. In cases where this Manual provides sufficiently detailed information, lower-level criticality safety procedures or manuals may incorporate such information directly or by reference to this Manual.

Nuclear criticality safety is administered at the Savannah River Site by assigning responsibilities for key nuclear criticality safety requirements and activities to appropriate WSRC organizational units. The WSRC President shall:

1. Establish the company-level policy for implementing the criticality safety requirements of the Department of Energy and make it known to all affected WSRC employees involved in operations with fissionable material through written company policies.

2. Accept overall responsibility for the criticality safety of operations.

3. Delegate the authority and assign the responsibility for the day-to-day criticality safety of operations to lower-level management. Specifically, it is the responsibility of the division level managers and division chief engineers having control of facilities and processes

involving fissionable materials to assure that the applicable requirements of this Manual are implemented within their division. Criticality safety should be assigned importance commensurate with the other aspects of division management.

4. Establish a system of WSRC committees whose purpose is to: monitor the site criticality safety program; provide guidance and consultation to facilities; work with facilities on criticality safety questions and issues; assist in identifying potential improvements in equipment and operations; and otherwise promote the site nuclear criticality safety program. Presently established committees include the:

   4.1 NCSRC, which reports to the TEMIC. The NCSRC is the highest level site criticality safety committee and is responsible for fostering criticality safety across the site and serving as an arbiter in resolving criticality safety technical and policy issues. The NCSRC annually reviews the activities of the Criticality Safety Committees and the NIM Committee, which report to the NCSRC.
   4.2 Criticality Safety Committees (CSCs) which report to the NCSRC. The CSCs are maintained in the Nuclear Materials Management Division (NMMD); High Level Waste Division (HLWD); and Savannah River Technology Center (SRTC) to assist in self-assessment, provide technical support, and perform inprocess review as necessary regarding fissionable material handling, storage, processing, and disposal activities from a criticality safety perspective.
   4.3 NIM Committee, which reports to the NCSRC. The NIM Committee reviews nuclear incident monitor design changes, troubleshooting, and maintenance, and establishes site standards for NIM wiring and mounting. However, the NIM Committee does not have the programmatic responsibility for the site-wide NIM program. As delegated by the Site Chief Engineer, the Chairman of the Nuclear Criticality Safety Review Committee has programmatic responsibility for the site-wide NIM program.

5. Delegate the responsibility to perform an Integrated Management Evaluation of the WSRC criticality safety program to the WSRC Site Chief Engineer.

The NMMD; Facilities Disposition Division; High Level Waste Division; Solid Waste Division; and Savannah River Technology Center (SRTC) (i.e., Division General Manager and/or Chief Engineer) shall:

1. Accept and implement criticality safety responsibilities as delegated by the WSRC President.
2. Accept the responsibility for criticality safety for facility operations under their control.
3. Ensure that applicable criticality safety standards and DOE Orders are applied in the design, modification, and operation of facilities under their control. Means to ensure that this requirement is met potentially and include the use of configuration control boards, design reviews, technical reviews, safety evaluations, operational readiness reviews (ORRs), self-assessments, and training, as applicable.
4. Ensure that the WSRC policy for nuclear criticality safety is implemented at the Division level. Means to ensure that this requirement is met potentially include the use of Design Reviews; Safety Analysis Reports; Double Contingency Analyses; Operational Safety Requirements; Technical Safety Requirements; Technical Standards; Nuclear Criticality Safety Evaluations; Nuclear Incident Monitors; Test Authorizations and Test Conclusions; designated evacuation routes; independent review and oversight; and training.
5. Ensure that an interface or matrix arrangement exists with WSMS/CRT to provide necessary criticality safety expertise to facility activities and operations.
6. Ensure that appropriate training is developed and provided to staff under their control in compliance with applicable DOE Orders. However, the specific responsibility to qualify criticality safety engineers, or others, in the use of criticality safety computer codes required by the Divisions is assigned to WSMS/CRT. Division criticality safety training shall be reviewed as part of self-assessments and Facility Evaluation Board (FEB) reviews.
7. Delegate criticality safety responsibilities to lower-level managers within the Division as necessary; however overall accountability for such responsibilities resides at the Division level. Lower-level managers, facility managers, and supervisors should be evaluated, as

applicable, with respect to their performance in criticality safety on an annual basis (noting criticality safety related occurrences, limit violations, commendable practices, etc.) in the same way that they are evaluated for other job responsibilities during their regular annual review.

8. Maintain a Division/Area level Criticality Safety Committee (CSC), as appropriate, and document the authority and responsibility of the committee in Division/Area manuals and/or committee charters. Alternatively, a Division/Area may arrange to utilize the CSC of another Division/Area. For example, SW uses the NMSS CSC. Such arrangements shall be documented. These committees report to the NCSRC.

The Environment, Safety, Health & Quality Assurance (ESH & QA) Division shall:

1. Accept and implement criticality safety responsibilities as delegated by the WSRC President.
2. Provide personnel familiar with the physics of criticality and with associated safety practices to participate in Facility Evaluation Boards and Operational Readiness Reviews.
3. Provide personnel with criticality safety experience to serve on Area Criticality Safety Committees, as requested.
4. Provide Radiological Control & Health Physics Department functions for the calibration, use, and control of sources used to perform NIM field response checks; and for the calibration, use and adequacy of dosimetry and field-survey instrumentation used in conjunction with NIM evacuations.
5. Include criticality safety among the functional areas for Facility Evaluation Board review or Operational Readiness Review of facilities having significant quantities of fissionable material, or facilities which have to periodically demonstrate that a significant quantity of fissionable material is not present (e.g., HLW tank farm). The level of detail of such reviews shall be commensurate with the number and type of criticality safety controls and limits employed, the complexity of operations, and the mass and form of fissionable material.

The Projects Engineering & Construction Division (PE&C) shall accept responsibilities in the area of criticality safety as assigned by the WSRC President and shall comply with this manual as it pertains to criticality safety aspects of design. PE&C shall comply with the criticality safety design requirements contained in DOE Order 420.1 (DOE 2000b), as well as criticality safety design requirements contained in this Manual. The Chairman of the NCSRC has the programmatic responsibility for the site-wide NIM program as delegated by the Site Chief Engineer. PE&C, through its WSMS affiliate, shall provide risk assessments of the potential for criticality accidents as necessary for Safety Analysis Reports, reliability analyses, and other analyses in support of nuclear criticality safety analyses, as requested. PE&C, through its WSMS affiliate, shall periodically provide personnel with criticality safety experience to serve on Area Criticality Safety Committees and assist the Nuclear Criticality Safety Review Committee, as requested. PE&C shall ensure that a criticality safety organization exists to provide criticality safety engineering services across the site. The present WSMS/CRT affiliate provides such services. The WSMS/CRT shall provide criticality safety engineering support on a matrix basis to divisions/areas having significant quantities of fissionable material and, on an as-needed basis, to divisions/areas that have to demonstrate that a significant quantity of fissionable material does not exist for their facilities. WSMS/CRT personnel matrixed to other WSRC divisions should be administratively independent of operations (to the extent practicable), and shall be familiar with the physics of criticality, facility operations, and associated safety practices. WSMS/CRT shall provide the necessary computational tools (including necessary documentation of such tools) and methods for use by the site criticality safety engineers. WSMS/CRT shall also provide staff for independent reviews of NCSEs, NSDSs, criticality safety limit documents, and the criticality safety sections of safety basis documents. WSMS/CRT shall be responsible for the development and maintenance of the *Nuclear Criticality Safety Methods Manual* (WSMS 2001). The WSMS/CRT shall provide for review of nuclear criticality safety evaluations (NCSEs), nuclear safety data sheets (NSDSs), and engineering calculations. These reviews shall be performed by personnel who were not involved in the preparation of such documents. WSMS/CRT shall review and approve all formal documents that establish or eliminate criticality safety limits, or establish that limits are not required.

The WSMS/CRT shall verify, validate, and provide criticality safety engineers with controlled neutronics codes (e.g., keff, buckling, kinetics calculations) and data sets for use on their own computer systems and/or maintain a set of such codes on centralized site computer systems, including established bias and bias uncertainties (or data by which bias and bias uncertainties can be constructed). The uncertainty in the bias is interpreted as a measure of both the accuracy of the calculation and the precision of the experimental data. It is assumed to also include the precision of the calculation, if the calculation is stochastic (notwithstanding that such precision can often be made as high as desired), and the accuracy of the experimental data if the experiment is a mock-up of a referenced system. Deterministic computer calculations are assumed to have very high precision, or should be made to have very high precision. Stochastic computer calculations should be forced to have appropriately high precision. An experiment that is not a mock up is exactly accurate by definition. The WSMS/CRT shall verify, validate, and provide criticality safety engineers with neutron/gamma transport (i.e., shielding) codes and associated nuclear data sets. The WSMS/CRT shall qualify criticality safety engineers, and others, in the use of neutronics codes and gamma/neutron transport codes. The WSMS/CRT shall perform NIM location analyses and document results. The WSMS/CRT shall also conduct analyses to determine the evacuation-zone boundaries and establish source terms for credible accidents as requested. For NIM location analysis and evacuation boundary-zone analysis, these calculations may be based on methods ranging from the use of empirical/analytical equations to sophisticated three-dimensional computer code analysis. For source term estimation, the analysis may range from comparison of credible scenarios for a given facility against documented accidents/excursions to the use of kinetics codes, as appropriate. The WSMS/CRT staff should maintain familiarity with current developments in nuclear criticality safety standards, guides, and codes. Knowledge of current information should be maintained. The WSMS/CRT staff should also consult with knowledgeable individuals to obtain technical information as needed. The WSMS/CRT shall initiate formal criticality safety documentation of recommended limits. The WSMS/CRT shall provide assistance in the development, and shall provide review and concurrence of operating procedures (including special procedures) and procedure changes affecting criticality safety (including emergency procedures).

The WSMS/CRT shall review and concur in design drawings, and drawing changes, affecting criticality safety. The WSMS/CRT shall provide technical assistance in equipment and process design, and shall review and concur in process and equipment changes affecting criticality safety, particularly passive engineered controls and active engineered controls. WSMS/CRT staff shall provide first-level independent evaluation of criticality safety control effectiveness. WSMS/CRT staff shall recommend to facility management that any operation be brought to a safe condition (e.g., stop movement or processing of fissionable material) which, in their judgement, does not have the required level of criticality safety. WSMS/CRT staff shall maintain familiarity with all operations within organizations for which criticality safety support is provided and which require criticality safety controls. Familiarity may be gained through a combination of reading, tours, training, inspections, calculations, and periodic assignment to a facility as appropriate. WSMS/CRT staff shall participate on the Division/Area CSCs for the purpose of assessing criticality safety practices, providing in-process review, and providing technical support as necessary. The WSMS/CRT shall provide technical information, as requested, for general and facility-specific criticality safety training curricula, and review such training periodically. The WSMS/CRT shall review the Fire Control Pre-Plan for each facility in which moderation control is a concern. The WSMS/CRT shall examine reports of procedural violations and other deficiencies for potential improvements of safety practices and procedural requirements, and report such potential improvements to management. The WSMS/CRT shall provide personnel familiar with the physics of criticality and with associated safety practices to prepare and maintain the site Criticality Safety Manual.

Facility Managers, whose facilities warrant criticality safety consideration and controls, shall:

1. Accept responsibility for the day-to-day criticality safety of their facility.
2. Accept responsibility for and administer a program for NIM system management within the facility, including but not limited to: establishment of sufficient NIM coverage for facility operations and NIM availability; control-room alarm systems, as necessary; posting of NIM evacuation routes; maintenance of a current copy of the NIM

location and associated boundary zone analysis in facility files; and timely generation of requests for additional NIM analysis (location of new or existing NIMs, source term development, bell location or boundary zone changes) as required by facility and process changes.

3. Accept responsibility for preparation and maintenance of specific criticality safety procedures, or procedures for facility operation which identify nuclear safety steps/controls, and oversee the use of and adherence to such procedures in day-to-day operations.
4. Maintain a program of staff training (see Table 2) in both the general and facility specific aspects of criticality safety as appropriate. This includes: 1) maintaining a program of NIM familiarization and evacuation instruction for non-WSRC visitors and WSRC personnel needing access, but not normally assigned, to the facility (e.g., FEB and ORR teams); and 2) providing any additional criticality safety training as deemed necessary for non-WSRC visitors and WSRC personnel needing access, but not normally assigned to the facility.
5. Provide appropriately-trained staff to determine when procedures (including special procedures), drawings, and design documents require criticality safety review, and ensure that such procedures, drawings, and design documents are forwarded to the WSMS/CRT for review.
6. Ensure that criticality safety is included in facility self-assessments. This includes responding to criticality safety related deficiencies or action items noted by the responsible CSC and the findings and corrective actions proposed by all other internal and external review groups.
7. Accept responsibility for compliance with applicable DOE Orders; Operational Safety Requirements (OSRs)/Technical Safety Requirements (TSRs); Technical Standards; and Nuclear Criticality Safety Supplements. WSMS/CRT personnel shall assist facilities in ensuring compliance.
8. Approve the response to action items noted by the CSC.
9. Ensure that criticality safety aspects of facility design, construction, and operation are covered by a documented safety analysis.
10. Ensure that the facility is covered by a documented double contingency analysis.
11. Accept responsibility for placing their facilities in a safe condition (e.g., stop movement or processing of fissionable material) where warranted by actual or indicated unsafe criticality conditions.

Table 2. Summary of criticality safety training requirements.

	General Training (4)	Facility Specific	Operational Training
Operators/Fissionable/Accountable Material Handler (7)	Appl	Appl	Appl
Supervisors (1)	Appl	Appl	Not/Appl
E&I/Maintenance Mechanics	Appl	Appl	Appl (5)
Construction (8)	Appl	Not/Appl	Not/Appl
Facility Support Engineers	Appl	Appl	Appl (2)
Facility Management (3)	Appl	Not/Appl	Not/Appl
Design/NIM/Nuclear Material Packaging Engineers; FEB/ORR Teams (6)	Appl	Not/Appl	Not/Appl

Notes:

1. Supervisor training involves additional material. Supervisors should be capable of providing operational training to operators as necessary.
2. Includes any engineer performing or reviewing activities in which fissionable material is involved or could be affected; otherwise at discretion of management.
3. As a minimum, facility managers designated level 3 or subordinate shall be trained.
4. Includes NIM alarm familiarization, evacuation route signs, NIM evacuation routes, expected employee response, and rally point identification.
5. This training may require facility-specific aspects of several facilities.
6. The type of training and any testing is at the discretion of the management of these individuals. Qualification is not necessary.
7. Operators not involved with fissionable material handling, or processes or equipment involving fissionable material, only require general training.
8. Topics need only be covered as applicable for a specific job activity. Training may be performed at tool box meetings, in which case testing is optional.

12. Ensure that passive engineered, active engineered, and/or administrative criticality safety controls and associated limits are in place and functioning satisfactorily.
13. Accept responsibility for the development of a facility fire safety plan which recognizes, to the extent necessary, both fire safety and criticality safety considerations.
14. Establish and maintain criticality safety postings for the facility and labeling of fissionable materials.
15. Delegate criticality safety responsibilities to lower-level facility supervision, as necessary; however, overall responsibility for facility criticality safety remains with the facility manager.

16. Ensure that NIM evacuation routes provide for timely facility evacuation, that facility changes do not unnecessarily impede or otherwise lengthen evacuation time, and that, to the extent practical, routes do not require personnel to approach potential sites of a criticality accident.
17. Provide monitoring and surveillance as necessary to detect and prevent inadvertent accumulations of fissionable materials in process equipment, storage areas, piping, and ventilation systems. If accumulations of fissionable materials are detected, corrective actions shall be taken as necessary, accounting for recommendations of the WSMS/CRT.
18. Provide other criticality safety features and administration as necessary to provide for the criticality safety of the facility, including facility access control.

Facility operators shall:

1. Be trained in criticality safety commensurate with their duties, which typically will consist of general and facility-specific training.
2. Be responsible for criticality safety of their own actions and the operating systems under their control.
3. Be familiar with and follow criticality safety procedures as written or adhere to all nuclear criticality safety steps in operating procedures related to their assignment, as applicable.
4. Stop performing a procedure if it does not correspond to the operating system upon which it is being performed, or if an unforeseen condition develops, and notify supervision.

First-line supervisors shall:

1. Accept responsibility for the criticality safety of operations under their control.
2. Be knowledgeable in those aspects of criticality safety relevant to operations under their control.
3. Ensure that criticality safety training is administered to personnel under their control and require that these personnel have an understanding of procedures and safety considerations necessary to perform their functions without undue risk. Records of training activities and verification of personnel understanding, shall be maintained.

4. Develop or participate in the development of written procedures applicable to the operations under their control. Maintenance of these procedures to reflect changes in operation shall be a continuing supervisory responsibility.
5. Verify compliance with criticality safety specifications for new or modified equipment prior to its use.
6. Require conformance with good safety practices including unambiguous identification of fissile materials and good housekeeping.
7. Obtain training and assistance as necessary from the WSMS/CRT.

Facilities in which criticality is a concern can be divided into three general categories:

1. Facilities that, because of the quantity of fissionable material(s) and the processes and operations involved, require controls to prevent criticality accidents and reduce the probability of a criticality accident to a level judged acceptable, although still credible, in the Documented Safety Analysis (DSA). Examples of such facilities include FB-Line, HB-Line, F-Canyon, and H-Canyon.

2. Facilities that, because of the quantity of fissionable material(s) and the processes and operations involved, require controls to prevent criticality accidents, but by implementing such controls, reduce the probability of a criticality accident to a level judged incredible. Examples of such facilities include SRTC, 772-F/1F, and 235-F.

3. Facilities in which, based on documented engineering judgement or comprehensive analysis, a criticality accident is determined to be incredible and controls are not required, but the as-designed process conditions and assumptions upon which incredibility is based, are maintained. An example of such a facility is the 242-16H evaporator. Training, procedures, change control, and periodic assessments are all important in maintaining the as-designed process conditions and assumptions.

Some facilities meet more than one category because of the diversity of operations taking place within the facility. If a criticality accident is inherently incredible based on the quantity of fissile material, form of the material, or the inherent physical or chemical nature of the process, then application of this Manual is not required. However, for facilities having more than 450 g of ^{239}Pu or 700 g of ^{235}U, a determination of inherent incredibility shall be documented by the facility, and WSMS/CRT concurrence obtained. Reviews of facility or process changes (e.g., Design Authority Technical Review (DATR) or USQE) shall evaluate the need for criticality safety requirements.

The general criticality safety control principles and practices are:

1. Criticality safety shall not be compromised for the sake of expediency, production, or economic pressure.
2. An assessment of a facility or equipment design shall be made as early as practical to determine if the potential for a criticality accident exists. When such potential exists, facility and equipment designs shall meet applicable DOE Orders, ANS-8 standards, and other regulations related to criticality safety.
3. Criticality calculations for undissolved reactor fuels and targets should normally be based on beginning of life (pre-irradiation) fissile concentrations and enrichments. However, pre-irradiation fissile concentrations may have to be exceeded when justified by in-growth. Similarly, credit for the depletion of fissile material (i.e., burn-up credit) may be used when justified by known exposure history and/or appropriate measurement techniques. For example, credit may be taken for fuel burn-up by establishing a maximum fuel unit reactivity and assuring that each fuel unit has a reactivity less than the established maximum as determined by a reactivity measurement or by analysis and verification of the exposure history of each fuel unit. Credit may be taken for neutron absorbing materials that are incorporated into the fuel itself (e.g., burnable poisons or fission products), in structures or equipment which are a part of or contain the fuel unit, or both if the continued presence of the absorber, its distribution, and its concentration can be verified.
4. Unless a specific nuclear criticality safety evaluation has been performed, storage of fissile materials should be consistent with the guidance provided in ANS-8.7 (ANS 1998b).
5. Criticality control of special actinide elements shall be consistent with ANS-8.15 (ANS 1981) unless specific nuclear criticality safety evaluations have been performed. Operations may be performed safely

by complying with the subcritical mass limits given in ANS-8.15 (ANS 1981) Tables 1 and 2. In PuO_2-H_2O mixtures, regardless of the H/Pu atomic ratio, a subcritical limit of 8 kg of Pu is valid provided that: 1) the Pu contains at least 67% ^{238}Pu; 2) the isotopic concentration of ^{241}Pu is less than ^{240}Pu; and 3) the surrounding materials, including nearby fissionable materials, can be shown to increase k_{eff} no more than enclosing the unit by a layer of water of unlimited thickness. If Table 4 of ANS-8.15 (ANS 1981) is used, 3 g ^{243}Cm or 30 g ^{247}Cm will be considered equivalent to 1 g ^{245}Cm.

6. Subcritical neutron multiplication measurements shall be consistent with ANS-8.6 "Safety in Conducting Subcritical Neutron Multiplication Measurements In Situ" (ANS 1983).
7. The implementation of, functioning, and employee familiarization with the NIM system shall be consistent with this Manual.
8. Design of processes or equipment should make proper operation convenient and misoperation either inconvenient or impossible.
9. Before beginning an operation involving fissionable materials, or changing an existing operation, it shall be determined that the entire process will be subcritical under both normal and credible abnormal operating conditions (e.g., control of spacing, mass, density, geometry, moderation, reflection, enrichment, absorbers). This determination shall be documented in one or more Nuclear Criticality Safety Evaluation(s) (NCSE). In addition, this determination shall be verified (reviewed) as part of the facility operational readiness review. Care shall be exercised to determine those conditions that result in the maximum effective neutron multiplication factor; that is, the influence of variations of the controlled parameters on the system multiplication factor shall be understood.
10. The basis for establishing subcriticality shall be identified for all significant conditions in each process or operation (i.e., all criticality safety parameters and their limits shall be specified). In the case of established facilities or operations, this may consist of references to existing NCSEs.
11. Process specifications shall incorporate margins to protect against uncertainties in process variables and against a limit being accidentally exceeded.
12. The following materials are exempt from criticality safety control if no other fissile material is handled in the same facility:

12.1 No more than 15 g of ^{241}Am, ^{243}Am, ^{244}Cm, or any fissile material having an atomic number less than 95. Similarly, no more than140 g of ^{238}Pu, or no more than 1100 g of ^{237}Np, are also exempt based on a mass reduction factor of 30. Note that 10 CFR71.53 (USNRC 2002) exempts 15 g of ^{233}U, ^{235}U, ^{239}Pu, or ^{241}Pu, which provides a safety factor of about 30, or more, in terms of mass.

12.2 No more than 1g of other Am, Cm, or Cf fissile isotopes. Note that as given in ANS-8.15 (ANS 1981), based on fissile masses, this provides a safety factor of approximately 20 for ^{242}Am and substantially more for other isotopes. These other isotopes rarely exist in relatively pure forms and are generally included as part of spent fuel.

Other items (e.g., sealed sources, etc.) may be defined as exempt material on a case-by-case basis by the cognizant CSC and documented. Documentation may be in the form of a letter from the CSC to the facility, should summarize the basis for criticality safety and any associated criticality safety restrictions, and should reference applicable engineering calculations. Material defined as exempt from criticality safety requirements based on this section does not preclude applicable nuclear material accountability requirements for such material.

13. Exempt quantities of fissionable materials may be transferred on site in approved packagings that have not been reviewed for criticality safety. If such materials are transported in bulk, the exempt material limit applies to the conveyance.
14. Process equipment used in areas from which immediate evacuation may be required should be so designed that leaving the equipment will not introduce significant risk.
15. Operations with fissile material may be performed safely by complying with any one of the limits given in ANS-8.1, sections 5.1, 5.2, 5.3, and 5.4 (and accompanying tables) for single units provided the conditions under which the limit applies are maintained. These limits shall be applied only when surrounding materials, including other nearby fissionable materials, can be shown to increase k_{eff} no more than does enclosing the unit by a contiguous layer of water of unlimited thickness. Alternatively, multi-parameter control using the limits contained in ANS-8.1 (ANS 1998a), sections 6.1 through 6.4, may be used

provided that surrounding materials can be shown to increase k_{eff} no more than thick water reflection.

16. Criticality control in designated moderator control areas shall be consistent with ANS-8.22 (ANS 1997b).

Criticality prevention shall be based upon the double contingency principle. When implementing the Double Contingency Principle, the contingencies (i.e., events dependent on chance, accident, or uncertain conditions) evaluated shall be independent of each other and not the result of common mode failures. Judgement is required in determining whether two accidental events are related and consequently whether they actually represent two contingencies or a single contingency. For example, exceeding storage limits and then flooding an area would constitute two independent events; however, fire followed by flooding from an automatic sprinkler system would be considered a single event. Single events beyond the Design Basis Accidents (DBAs) are outside the scope of double contingency requirements. A double contingency analysis shall be included as part of NCSEs which establish limits/controls for new or modified processes, pieces of equipment, storage, and transport involving fissionable material. This analysis may be qualitative in approach. The overall double contingency analysis for a facility shall be documented in an NCSE or NCSASR. Double contingency defenses shall be highly reliable. Engineering judgement may be the basis of making a determination that a control is highly reliable. Each double contingency defense should be capable of maintaining the minimum subcritical margin. Criticality safety parameters which are not controlled to some limit shall be assumed to take the most reactive credible values when evaluating the sufficiency of the double contingency defenses. If double contingency defenses are not independent, areas of overlap shall be identified (e.g., common power supplies, common methods of calibration, common components), and steps taken to remove common cause failure dependencies to the extent practical. If overlap of double contingency defenses cannot be eliminated, the overlap may be acceptable if the common features can be shown to fail in a manner that maintains the Minimum Subcritical Margin. Alternatively, overlapping double contingency defenses may be acceptable if shown to be sufficiently reliable.

To the extent practicable, double contingency defenses should employ passive engineered controls over active engineered controls over administrative controls. All reasonable efforts shall be directed toward avoiding the use of administrative controls only as the sole defenses to a criticality accident. If double contingency protection cannot be provided, an exemption from S/RID shall be obtained. Process and operational assumptions and conditions used in criticality accident analysis need to be periodically reviewed concurrent with facility DSA updates to assure their continued applicability and validity. Potential items to review are: 1) Normal process design; 2) Normal operating practices, operation, and expectations; 3) Proper function of design features; 4) Devices that monitor for and prevent an initiating event; and 5) Administrative NCS programs, NCS limits, and NCS controls. Postulated accident scenarios shall obey the laws of chemistry and physics, or alternatively, laws of chemistry and physics may be used as the basis for demonstration that an accident cannot occur. Procedure review, design review, control of modifications, control of maintenance, DATR, USQ process, review by a cognizant facility criticality safety engineer, and periodic assessments may be credited as protecting the process and operational conditions and assumptions related to criticality safety. These programs protect and maintain: normal process design; normal operating practices, control and expectations; and proper function of design features. When quantitative techniques are used to estimate the frequency of a criticality event, and the failure of an instrumented control or design feature is used as an initiating event (or as part of an accident sequence), then the instrumented control or design feature shall be implemented, tested, and maintained such that the estimated failure frequency remains valid. When quantitative techniques are used to estimate the frequency of an initiating event (or the frequency of an event contributing to an accident sequence) and the event is not dependent on controls or design features, then there are no requirements except to evaluate any process or operational changes that may invalidate or change the frequency. To the extent practical, the estimated frequency of initiating events, or events contributing to an accident sequence, should make use of operating experience.

The means of control for criticality safety refer to how control is achieved in design and operating terms using the criticality safety parameters. The three basic means of control are:

1. Passive Engineered Means of Control. Passive engineered control is the highest ranked means of criticality control, involving fixed, passive design features, or devices rather than moving parts (for example, a nuclear safety blank). Such means of control provide high reliability, a broad range of coverage (i.e., cover many potential criticality accident scenarios), and require little operational support to maintain effectiveness. Human intervention is not required. When practical, advantage is taken of natural forces, such as gravity, rather than electrical, mechanical, or hydraulic action.

2. Active Engineered Means of Control. Active engineered control is a means of control, intermediate in rank, involving add-on, active hardware (i.e., electrical, mechanical, hydraulic) that protect against criticality. These devices act by sensing a parameter important to criticality safety and providing automatic action to secure the system to a safe condition without the necessity of human intervention. However, active engineered devices are subject to random failure during operation and human error during maintenance and calibration. High quality/low failure rate equipment should be selected in all cases. Failsafe designs should be employed, if possible, and equipment failure should be easily and quickly detectable. The use of redundant systems should be considered as a means of dealing with unavailability.

3. Administrative Control. Administrative control is a means of control which relies on the judgement, training, and responsibility of people for implementation. Administrative means of control may be action, caution, or verification steps in a procedure, or steps in a surveillance program. Because these means of control are human based, and therefore subject to error in application, they are generally regarded as the least desirable means of criticality control. In some instances, however, reliance must be placed on this means of control, at least in part. Administrative control is often enhanced by the use of equipment which alerts operating personnel of some condition (e.g., pressure gauge, temperature, level indicator).

Under both normal and credible accident conditions, control of one or more of the following parameters shall be incorporated into the design, operation, and administration of a fissionable material facility to the extent necessary:

1. Geometry Control. Passive engineered controls such as geometry control shall be considered as a preferred control method. However, the facility double contingency analysis shall justify the final controls selected for use. Full advantage may be taken of any nuclear characteristics of the process materials and equipment. All dimensions and nuclear properties upon which reliance is placed shall be verified prior to beginning operation and continuing control shall be exercised over such properties and dimensions. The practicality of using geometry control involves: the type of equipment needed; the process flow rates and volumes; the inherent complexity of the process and economies of scale; maintenance requirements; and the viability of other types of criticality safety controls which lend themselves to the process. Because each system and facility may be unique, decisions must be made on a case-by-case basis and approved by the WSMS/CRT. Process systems shall be designed to prevent the carryover or inadvertent transfer of fissionable material capable of causing a criticality accident from geometrically favorable portions of a facility to other areas not having geometry control. In this regard, a system of positive control and backflow prevention such as air gaps shall be used to prevent the inadvertent transfer of fissionable material. When using geometry control, allowances shall be made for corrosion, distortion, erosion, and manufacturing tolerances. If distortion is a potential problem, steps shall be taken to prevent it (e.g., pressure relief valves, internal stay-bolts, rupture discs, heavier wall thickness, external bracing, and the use of more inherently stable geometries). In addition, the margin of safety shall be maintained under the worst credible geometry conditions. Criticality safety controls shall include provisions for periodic evaluation by an inspection program or use of corrosion specimens, if credible corrosion or erosion could change the geometry in a system which depends on geometry for criticality safety. The design of heating or cooling jackets shall include consideration of leaks of fissionable material into such jackets. Appropriate measures shall be taken to preclude accumulations of fissionable material in heating or cooling jackets. If accumulations occur in heating/cooling jackets, they shall not result in an unacceptable reduction in the subcritical margin. Sumps shall be designed such that criticality safety is ensured if a credible mechanism exists for accumulating fissionable material in the sump. Where applicable, floor drains shall be designed to preclude the

accumulation of fissionable material in traps and piping which could cause a criticality accident.

2. Spacing (Interaction) Control. Individual items of equipment and containers holding fissionable materials, when arranged in a group, in storage, or when being transferred within a nuclear facility or between facilities onsite, shall be spaced so that the entire array is subcritical for all normal or credible abnormal conditions that pertain to the nuclear facility or transfer. Storage racks whose purpose is to maintain spacing control shall be designed to maintain their integrity during and following design basis accidents (provided that the facility itself is designed to withstand DBAs) and shall be noncombustible. Spacing of storage units may be maintained by the use of birdcage fixtures, covered metal cans, or physical barriers on shelves subject to approval by an NCSE.

3. Neutron Absorber (Poison) Control. Neutron absorbing materials, such as cadmium, boron, and gadolinium may be used to make equipment and processes safe, provided available data confirm their suitability and assure their presence and reliability. Care should be exercised in the use of solutions of neutron absorbers because of the controls required to assure their continued effectiveness. The use of borosilicate-glass Raschig rings for packed vessels shall be consistent with ANS-8.5 (ANS 1996a). Use of representative samples, such as corrosion coupons, to verify the continued presence of a fixed poison shall require approved documentation which demonstrates that the samples actually represent the poison system. The use of borosilicate-glass Raschig rings for applications other than packed vessels shall be based on a documented NCSE and meet the requirements of this manual for fixed neutron absorbers. When installing Raschig rings in tanks that were previously used for fissile material, assurance should be required that the vessel contains less than the single parameter mass limit for the nuclide(s) of interest prior to partially filling the vessel with water or other appropriate liquid used during ring installation. When Raschig rings are used as the principle method of criticality control for solutions of ^{233}U, uranium enriched to greater than 5.0 wt% in ^{235}U, or plutonium in tanks, a fissile material balance should be performed with a frequency sufficient to preclude unsafe accumulations of fissile material in the tank and to maintain an

adequate subcritical margin of safety. When soluble absorbers are relied upon for criticality control, the minimum soluble poison concentration should be sufficient to ensure the margin of safety (i.e., maintain k-safe) when assuming all other variables are at their worst-case (credible) or Limiting Condition for Operation (LCO) values. At SRS, one recommended method is that two independent sample analyses or methods of determining the operating concentration of a soluble poison shall be provided to confirm that the poison concentration limit is satisfied. Fixed neutron absorbers shall be designed to maintain their required geometrical relationship with fissionable materials during the intended operating life. A means of verification shall be provided to determine that the design, safety, and operating requirements are met for all fixed neutron absorber system components. Requirements for in-service verification, and the frequency and extent of such inspections, shall be considered during the design of the fixed neutron absorber system. The inspection and verification plan shall conform to facility quality assurance requirements and shall consider associated safety analyses, the environment in which the absorbers are placed, and the absorber material properties. The inspection and verification plan shall cover material acquisition, manufacturing, installation, operation, and maintenance of the fixed absorber system. Any action as a result of the inspection plan shall not compromise the criticality safety of the system incorporating the absorbers. Inspection and verification shall be documented, and records maintained for the operating life of the facility. Verification of fixed neutron absorber material elemental and/or isotopic composition shall be obtained prior to use. Similarly, fixed neutron absorber system components shall be verified to conform to design drawings and specifications before installation; proper installation shall be verified before use. In-service verification methods used to measure the continued presence and effectiveness of a fixed neutron absorber may include neutronic techniques, other non-destructive techniques, or chemical testing methods. Analytical instruments, radiation sources, and calibration slabs, used to verify fixed neutron absorber properties shall be calibrated to NIST traceable standards or approved by the local CSC. The design of fixed neutron absorbers shall include assessment of the operating environment. Degradation of the fixed neutron absorber or materials to protect the neutron absorber due to chemical, physical, radiological, or mechanical impacts shall be assessed and protected against, and adequate allowances made during the design process. A fixed neutron absorber shall maintain its designed neutron absorption capability throughout its intended, operating lifetime, including all credible conditions of neutron moderation and reflection. Radiation effects on the neutron absorber system over its expected life shall be evaluated. Use of fixed neutron absorbers shall include allowances for process material variations, uncertainties in absorber density and distribution, and uncertainties in nuclear properties (e.g., absorption cross section) of the neutron absorber. The impact on the criticality safety of the system incorporating fixed absorbers shall include consideration of such uncertainties. Fixed neutron absorbers shall be designed such that criticality safety is not compromised during normal and credible abnormal operational or external (natural phenomena) events. A fixed neutron absorber system shall be designed to prevent inadvertent removal, displacement, or alteration of its component parts. The design of equipment and facilities incorporating fixed neutron absorbers shall include human factors engineering considerations for installation, operation, and maintenance of such absorbers. The requirements associated with operations, nuclear materials accountability, and other such disciplines shall be considered in the design of fixed neutron absorber systems. Any event found to subject a fixed neutron absorber system to physical or chemical conditions outside the design envelope shall require the reassessment of the system prior to resumption of operations. Calculation methods used to evaluate fixed neutron absorber effectiveness shall include consideration of localized neutron flux depression associated with such absorbers. Data from operations and maintenance shall be used to determine if fixed neutron absorber system performance conforms to initial design expectations and safety evaluations. Results of in-service verification shall be evaluated and corrective action taken as necessary.

4. Concentration (Density) Control. Systems that use concentration/density control shall meet established concentration/density criteria. These criteria may be found in criticality safety guides and handbooks. When reliance is placed upon concentration control alone for criticality safety, the concentration of fissile isotopes in uniform aqueous solutions shall not exceed the single parameter limits for aqueous solutions reflected by an effectively infinite thickness of water. Two

independent sample analyses or methods of determining the operating concentration shall be provided to confirm that a concentration limit is satisfied. Tanks containing fissile solutions shall be periodically monitored for evaporation and makeup solution added as necessary to prevent unsafe concentration. The frequency of this monitoring shall be determined on a case-by-case basis based on existing conditions. The chemical composition of makeup material shall be such as to prevent unwanted precipitation or phase changes.

5. Moderation and Reflection Control. For operations in which criticality safety depends upon control of neutron moderation, there shall be assurance (verification) that the prescribed extent of moderation remains unchanged or that, if a credible change occurs, the reactivity of the system remains below acceptable subcritical limits. Such assurance shall include consideration of all credible accidents involving any moderator or combination of moderators. Interstitial moderation shall be considered whenever such moderation is credible (e.g., low-density water due to water sprays, snow). If moderators or reflectors more effective than water may be present, their effects shall be considered. Facility fire protection programs shall establish guidelines for fire fighting within, or adjacent to, moderation controlled areas, as applicable. Criticality safety limits shall be based on worst credible reflection. Both normal operating conditions and credible accident conditions shall be considered. In general, reflection controls based on limiting personnel access to a system should not be relied upon to maintain criticality safety. However, in those cases where reflection controls are needed which are based on limiting personnel access to a system, the WSMS/CRT shall review, approve, and monitor such limits on a case-by-case basis. As required by an NCSE, enclosures and systems that penetrate a moderator control area (for example, material transport and transfer systems) should be minimized, but shall be designed and/or operated such that it is purposefully difficult to add fissionable and moderating material in excess of specified limits. Alternatively, appropriate administrative controls may be used to limit the unwanted addition of fissionable and moderating material to enclosures and systems that penetrate a moderator control area (for example, material transport and transfer systems) if design constraints do not achieve such purposes.

6. Mass and Volume Control. For operations depending upon mass controls, where the contained volume does not automatically limit the contents to a safe mass or less, the possibility of multiple batching or generally exceeding mass limits shall be evaluated. In areas where multiple batching is credible, minimum safety margins shall include such considerations. In addition, double batching shall be considered in storage safety analyses and in the establishment of operating procedures. If the calculation of a mass limit assumes fissile material in solution, it shall be shown that the change to a more reactive state due to precipitation or transfer to a second phase is sufficiently unlikely before controlling to that limit may be considered a contingency defense. Volumetric safety limits shall be based on the minimum volume required to sustain a neutron chain reaction for the geometric shape under consideration. The minimum critical volume shall be that associated with the worst credible process conditions that may exist within the system, including consideration of system interactions with other process systems and the environment. The openings for mass-controlled drain systems shall be covered when not in use.

7. Enrichment Control. For every operation, criticality safety limits shall be based upon the most reactive enrichment or isotopic composition that is credible. Documentation shall exist to establish the bases for such compositions and associated limits.

8. Temperature Control. In general, the lower the temperature, the lower the minimum critical mass. Such considerations may become important for items like cold traps which operate at temperatures substantially below the freezing point of water. If the temperature is substantially different from the temperature which is generally used as the basis for neutron cross sections for calculations, then changes to the cross sections should be considered.

CRITICALITY SAFETY TRAINING

It is WSRC (2002b) policy to provide its employees with the training and development opportunities necessary for them to complete their work assignments in a safe, effective, and quality manner. In the area of criticality safety training, the training policy supports the WSRC policy that all reasonable effort shall be taken to reduce or eliminate the potential for, and consequences of, a criticality accident. Criticality safety training is generally separated into three types of training depending on the employee's duties. The three types of training are: general training; facility-specific training; and operational training. General and facility-specific training are typically based on lectures, films, or video tapes and usually also require written examinations based on these lectures. In some cases, general and facility specific training may be combined. Operational training may consist of procedure walkdowns or may be included as part of an oral board examination. Employees receiving general and facility-specific criticality safety training shall also receive NIM evacuation alarm familiarization/response training as part of their overall criticality safety training and may be required to participate in NIM evacuation drills. However, there are also groups of WSRC employees who require training only in NIM evacuation alarm familiarization/response, some of whom may also be required to participate in NIM evacuation drills because of their work assignment. The type, depth, and emphasis of criticality safety training is dependent upon the employee's duties and position.

Divisions having facilities in which the requirements of this Manual are applicable shall each develop criticality safety training programs for their operating facility personnel. This criticality safety training shall be for facility operators; supervisors; fissionable material handlers; accountable material handlers; maintenance mechanics; engineers providing technical or process support; fissionable materials custodians; construction; facility management; lab technicians; and all other personnel whose job assignment requires them to be in an area where controls are instituted to ensure criticality safety. Divisions may work together to develop criticality safety training programs and/or train personnel. A summary of criticality safety training requirements is provided in Table 2. For those facilities in which all criticality accident scenarios have been judged incredible such that criticality safety controls are not required, but process conditions and assumptions important to the incredibility argument are to be maintained, training may be tailored to address only those items required to maintain incredibility. Such training needs to be given to those personnel directly involved in such activities, including level 3 and subordinate managers. Other groups requiring training include WSMS/CRT criticality safety engineers/specialists; PE&C design engineers; FEB and ORR personnel; NIM engineers; and nuclear packaging engineers.

Consideration shall be given to the nature of a person's assignment. Those persons whose responsibilities are managerial shall be provided training which includes information on how to manage criticality safety concerns. Those persons whose responsibilities are technical or operational shall be provided training which includes information related to the development of technical and operational competence. The training of supervisors shall be of increased depth as compared to operators/fissile material handlers. Each division shall document the criticality safety qualifications of each job category (education, experience, training, and examination) of staff identified in Table 2 (except for design engineers, FEB/ORR teams, site NIM engineers, and nuclear packaging engineers). Qualification of such persons shall be valid for a maximum of two years (unless revoked for cause), at which time personnel shall be requalified.

Personnel in each of the job categories listed in Table 2 (except for design engineers, FEB or ORR members, site NIM engineers, and nuclear packaging engineers) shall be qualified with respect to criticality safety training. Satisfactory completion of qualifications shall be verified by facility and/or company organizations designated as responsible for criticality safety training. Facility managers, level three and subordinate, shall complete general criticality safety training. Engineers/scientists directly assigned to a facility, and responsible for the operation or maintenance of processes, laboratories or equipment involving fissionable material in the facility, shall complete both general and facility-specific criticality safety training. This training shall be conducted biennially and shall be documented. Facility-specific training for particular levels of managers may be required at the discretion of Divisions/Areas/Facilities. Request for deviations from this requirement shall be reviewed by the cognizant CSC. Engineers who do not work directly in or for facilities may also require some form of criticality safety training. Design engineers (PE&C)

responsible for the design of facilities, processes, or equipment involving fissionable material should receive biennial criticality safety training. The content of PE&C design engineer training will be determined by PE&C management. FEB and ORR team members assigned the functional area of criticality safety should receive general training as determined necessary by their respective management. The content of, and need for, site NIM engineer general training shall be determined by, and at the discretion of, the NIM committee chairman. Nuclear packaging engineer training content and need shall be determined by, and at the discretion of, their management. Training for any of the groups mentioned may consist of reading assignments of appropriate material, lectures, and short courses, as necessary to meet needs. Testing and qualification shall not be required for design engineers, FEB and ORR teams, site NIM engineers, or nuclear packaging engineers and is left to the discretion of their respective management.

For those personnel identified in Table 2 (except as noted in footnote 8 in Table 2), either written, or written and oral (operational) examinations shall be prepared and administered for the training indicated in the Table. The written examinations may be divided into two separate training sessions: general criticality safety training, and facility-specific criticality safety training, or combined training. A minimum written and oral examination grade shall be established to distinguish between pass and fail. All incorrect answers to written examination questions shall be reviewed with the instructor. For employees requiring operational training, an oral (operational) examination shall also be provided. Failure of the general, facility-specific, or oral examinations shall preclude an individual from working on assignments directly involving criticality safety. Persons who fail written or operational tests may requalify by attending another training session, retesting, and achieving a passing score on the test.

General criticality safety training shall include, but not be limited to:

1. WSRC policy that all reasonable effort shall be taken to eliminate the potential for, and consequences of, a criticality accident
2. Historical information and descriptions of criticality accidents
3. Elements of atomic structure
4. Elements of neutron fission, absorption, and scattering reactions

5. Parameters affecting criticality, typical control methods and limits, and examples of passive engineered, active engineered, and administrative control (as applicable)
6. Characteristics of a criticality accident (e.g., fission burst, reburst, fission products, energy)
7. Radiation given off during a criticality accident
8. Health effects of acute radiation
9. Emergency actions (including NIM alarm, correct NIM alarm response, and evacuation routes)
10. Operating by procedure, criticality safety procedures, or nuclear safety steps in operating procedures (as applicable), and making changes to procedures/processes/equipment
11. Individual employee responsibility
12. Posting and labeling concepts, as applicable
13. Applicable sections of general Division/Area criticality manuals or procedures
14. Importance of configuration control.

Facility-specific criticality safety training shall include, but not be limited to, important job-specific aspects of:

1. Process flow sheets of systems having criticality safety significance, as related to specific work assignment, as applicable
2. Criticality safety control methods such as mass control, moderation control, and concentration control, utilized in the facility
3. Passive and/or active engineered equipment used for criticality safety, including its location, function, reliability, maintenance, and surveillance requirements, as applicable
4. Procedures/administrative controls used for criticality safety
5. Review of facility Safety Analysis Report and/or BIO criticality accident analysis
6. Fissionable material packaging and movement
7. Criticality Safety Operating Limits, Limiting Conditions of Operations, Limiting Control Settings, and other criticality safety related limits
8. Operational Safety Requirements, Technical Safety Requirements, Technical Standards
9. Criticality vs. fire safety (including important precautions, as applicable)

10. Nuclear Criticality Safety Supplements and Nuclear Safety Data Sheets (as applicable)
11. Operating characteristics of the facility as they pertain to criticality safety
12. Understanding and complying with moderator limits and controls.

Operational examinations may be conducted as a one-on-one walkthrough (e.g., procedure walkdown) or by an oral board/committee at the discretion of the facility. The operational aspects of criticality safety may be included as part of a larger operational evaluation. In any case, familiarity with the criticality safety aspects of facility operations as contained in facility procedures applicable to the candidate's job shall be demonstrated. Documentation of the candidate's operational examination shall be maintained. A minimum examination grade shall be established to distinguish between pass and fail. All incorrect answers shall be reviewed with the instructor. The operational examination may be conducted and documented procedure-by-procedure or by other convenient form.

Reexamination for qualified persons as identified in Table 2 shall include criticality safety material with which proficiency is expected. The minimum examination grade for passing shall be the same as the one established for regular testing. All incorrect answers shall be reviewed with the instructor. Persons identified in Table 2 shall not be allowed to function as qualified in their position if they have not completed criticality safety requalification within two years (defined as the end of calendar year in which requalification is due). If such person fails the requalification examination(s), or shows serious job performance deficiencies which indicate that he/she may perform in an unsafe manner, the person shall be reassigned to activities not requiring criticality safety qualification. Requalification in criticality safety may be regained upon completion of requalification training and reexamination. In addition, for personnel working in facilities in which fissionable material is handled, stored, or processed, requalification shall include consideration of an individual's operating performance with respect to criticality safety during the past qualification period. This review may be made by line management, a committee, or by a person designated by management. When a qualified person at a facility in which fissionable material is handled, stored, or processed, has been absent from qualification duties for greater than 3 months, but less than 12 months, selected retraining and reexamination shall be given as deemed necessary (e.g., because of procedure or equipment changes) by facility management or the area CSC. In such cases, the qualification base date remains the same as it was prior to the absence. However, if the absence is greater than 12 months, retraining and reexamination shall be required. In such cases, a new qualification date for criticality safety qualification shall be established. For facilities in which fissionable material is handled, stored, or processed, facility management shall develop procedures which define on-the-job requirements and frequency (i.e., "x" hours per month) necessary to maintain qualification with respect to criticality safety. The FEB and/or area CSC should periodically review such requirements and frequencies and make management aware of potential problems. If such requirements are not met, qualification shall be suspended and the person reassigned to duties not requiring qualification. Persons may regain qualification by performing their duties under the direct supervision of a qualified person in accordance with established procedures. If a facility is infrequently operated, qualification shall be reinstated prior to facility operation by administering written and operational examinations as necessary to ensure adequate operational knowledge. An extension of qualification may be granted to persons on a case-by-case basis in order to support operational and schedule commitments. A procedure shall be developed by facility management for this purpose. The area CSC may review such procedures for adequacy. Facility management should inform the area CSC of such proposed extensions and the CSC should review the proposed extensions as part of their technical support function.

The WSMS/CRT should provide technical information as requested for criticality safety training for divisions/areas having significant quantities of fissionable material, and should review training periodically. In addition, the Division/Area CSCs should review the criticality safety training at their discretion. WSMS/CRT shall maintain a training and qualification program for its nuclear criticality safety staff using a graded approach and be based on the duties and responsibilities of the individual members. For each level of staff, the program should include an appropriate combination of education, on-the-job experience, qualification to perform criticality safety calculations, general criticality safety training (e.g., on-site or off-site courses or classes), reading assignments, and facility specific training. Reentry Team members shall receive training on the use of radiation monitoring instrumentation of the type used for reentry, including instrumentation sufficient to detect a reburst; methods of locating the source of a critical excursion; appropriate anti-contamination clothing requirements; potential methods of disarming a critical excursion (if approved as part of the recovery plan); precautions before changing process conditions; and historical criticality accidents. Each Division/Area/Facility shall develop and implement procedures that specify requirements for the maintenance of criticality safety training and qualification records. In the case of design engineers, FEB and ORR teams, site NIM engineers, and nuclear packaging engineers, a specific procedure may not be necessary as long as training is documented and maintained in section/department files. Documentation of training for criticality safety engineers/specialists shall be maintained.

General Employee Training should include identification of the nuclear incident monitoring (NIM) system bell alarm; familiarization with the sound of the alarm; instruction on the requirement for immediate evacuation if in an area in which a NIM alarm sounds; identification of a NIM evacuation route sign; and a brief explanation of why a NIM alarm is necessary. All employees whose regular work assignment is in a facility within a NIM evacuation zone shall receive biennial NIM evacuation training including the: requirement for immediate evacuation any time the alarm sounds; sound of the alarm signal (and location and appearance of any associated lights); identification of NIM evacuation route signs, facility-specific evacuation routes, rally point location; and the reason for having a NIM alarm system. For those employees identified in Table 2, NIM training shall be included as part of their general criticality safety training. For those employees not identified in table 2 (e.g. clerical staff, or other technical or professional staff not having responsibilities involving fissionable material or processes or equipment containing fissionable material) whose regular work assignment is in a facility within a NIM evacuation zone, a separate, documented, biennial NIM evacuation training session is acceptable and testing is not required.

Before placing a new NIM system in operation, all employees normally working in the area shall be acquainted with the signal by actual demonstration at their work locations. Evacuation is required. To maintain familiarization and acquaint new employees and transferees into the area, the NIM alarm signal should be sounded periodically during working hours after notifying all concerned. Alarm familiarization activities shall include non-regular shift employees. Evacuation is required any time the bells are sounded. Annual evacuation drills constitute one aspect of employee alarm familiarization; but it is not mandatory that each facility employee participate in an annual drill. Semi-annual alarm operability and annual audibility testing affords other opportunities for facility employees to hear the alarm bell. Other alarm familiarization training includes information from CAT, Radworker, criticality safety training, and a recorded phone message. Each facility (including all areas within the evacuation zone) having a NIM alarm system shall conduct a minimum of one NIM evacuation drill per year. At the discretion of the facility manager, and in combination with other NIM associated training, additional drills may be performed to ensure that facility personnel understand the proper response to a NIM alarm. NIM evacuation drills shall be preceded by written notice, posted signs, or voice announcement. Surprise-test evacuations shall not be conducted. All WSRC employees within the evacuation zone at the time of the drill shall participate regardless of their normal assignment or normal location.

Untrained non-WSRC personnel visiting an area covered by a NIM system should be familiarized with the evacuation alarm signal and instructed in the proper response to a NIM alarm (and personally escorted while in the facility as necessary). Such non-WSRC personnel may be trained by attending GET/CAT class, the use of training material available at the facility visited, or by use of the recorded phone message. Persons not normally assigned to a facility having greater than exempt quantities of fissionable material shall also be instructed to refrain from any actions involving the handling, movement, processing, or storage of fissionable material within the facility. Facility-specific NIM evacuation route instructions for WSRC personnel needing access, but not normally assigned to a facility (e.g., FEB and ORR teams) should be maintained in all facilities equipped with NIMs. Facilities having NIMs should also provide any additional criticality safety training as deemed necessary for such WSRC personnel.

Peer Review Criteria, Findings, and Recommendations of the Review Panel

PEER REVIEW CRITERIA AND FINDINGS OF THE RP

The findings of the Review Panel (RP) with respect to the review criteria are as follows:

Criterion 1

Are the Nuclear Safety Program functions adequately defined with responsibilities and interfaces clearly understood?

Finding 1 of the RP

The Nuclear Safety Program functions are substantially defined in a number of documents such as: the *Westinghouse Savannah River Company Nuclear Criticality Safety Manual* (WSRC 2002b); Procedure 1.04, *Potential Inadequacy in Safety Analysis Declaration Process* and Procedure 1.05, *Nuclear Facility Unreviewed Safety Questions* of the *Facility Safety Document Manual* (WSRC 2001); and the *Conduct of Engineering and Technical Support Manual* (WSRC 2002h). Moreover, based on the information provided to the RP during the oral presentation, it appears that the Nuclear Safety Program functions and responsibilities are clearly understood by the technical staff. The interfaces could be better communicated and understood if there were a single document identifying all the elements of the whole Nuclear Safety Program.

The objective of the Nuclear Safety Programs at Savannah River Site (SRS) is to manage SRS in a manner which ensures that there is no undue risk of nuclear and process accidents that could adversely affect the health or safety of employees, visitors, members of the general public, or the environment. For all activities, the continued assurance of the capability and capacity for safe operations remains paramount to protect facilities and the environment from unacceptable risks. These programs apply to all personnel engaged in nuclear and process-related activities, and are in effect during design; construction; normal and abnormal operations; maintenance; modifications; surveillance; transition; deactivation; decontamination; and decommissioning. The intent is not only to prevent accidents, but to make provisions to limit consequences should accidents occur.

The Nuclear Safety Program is well-established and mature. The SRS management is commended for recognizing the need for emphasis on the integration of the safety functions. The Project Team presented to the RP the Disciplined Conduct of Projects (DCOP) Process and the Consolidated Hazard Analysis Process, which are part of the implementation of the Integrated Safety Management System (ISMS) for nuclear activities and projects.

The ISMS was graphically presented by the Project Team as a flowchart identifying activities and documents within the Nuclear Safety Program that could serve as a road map for the development of a single document consolidating the Nuclear Safety Program requirements.

The Project Team presented a recently-revised organizational chart that illustrates the consolidation of all nuclear safety management responsibilities at SRS under a single manager—which is a positive step. The effectiveness of this reorganization cannot be assessed because insufficient experience has been accumulated. Timely evaluation of the effectiveness of this new organization is necessary.

Criterion 2

Is each Nuclear Safety Program function assigned to an organization having the appropriate authority and accountability to ensure proper conduct?

Finding 2 of the RP

Based on the information provided to the RP during the oral presentation, it appears that the Nuclear Safety Program functions are assigned to organizations having the appropriate authority and accountability to ensure proper conduct. The details regarding the specific organizations and positions are identified in several documents such as: the *Westinghouse Savannah River Company Nuclear Criticality Safety Manual* (WSRC 2002b); Procedure 1.04, *Potential Inadequacy in Safety Analysis Declaration Process* and Procedure 1.05, *Nuclear Facility Unreviewed Safety Questions* of the *Facility Safety Document Manual* (WSRC 2001); and the *Conduct of Engineering and Technical Support*

(WSRC 2002h) Manual. All personnel engaged in nuclear and process-related activities, including all contractors, have to abide by the same requirements during design; construction; normal and abnormal operations; maintenance; modifications; surveillance; transition; deactivation; decontamination; and decommissioning.

The procedures identified above appear to be complex and cumbersome. The Project Team stated that it is working on the Safety Analysis Documentation Improvement Strategy (SADIS) to streamline and better integrate the SRS procedures, including those that are safety-related. The completion of this activity is important for, and could enhance, the safety performance of the SRS.

The Project Team stated that the Quality Assurance (QA) Department and the Facility Evaluation Board have completed audits and assessments indicating the satisfactory conduct of the Nuclear Safety Program.

The Project Team reported initial resistance of some managers that led to some problems in projects such as the Highly Enriched Uranium Blendown Project. Early establishment of the integrated Project Team—as required by DOE Order 413.3, *Program and Project Management for the Acquisition of Capital Assets* (DOE 2000c)—is likely to ensure better communication among project participants, and thus will reduce the potential occurrence of similar problems. The lesson learned is that integration of support functions with the facility operations needs continuing attention. Inclusion of the safety-related responsibilities in the personnel performance evaluation system would strengthen an individual's accountability and thus facilitate better implementation of the DCOP process. Team work is fundamental to most effectively attain and maintain the essential level of safety and mission commitments.

Criterion 3

Does the training for newly-assigned engineers develop the required nuclear safety-related work process (DOE Savannah River Site) knowledge and skills? In particular, is this training appropriate for engineers new to SRS?

Finding 3 of the RP

The information provided to the RP during the oral presentation was specific to the engineers assigned to High-Level Waste organizations. It appears that the training for these newly-assigned engineers systematically develops the required nuclear safety-related work process knowledge and skills. However, the Project Team acknowledged that this high quality program is not uniformly implemented across the site.

This training program is appropriate for engineers new to SRS because it focuses on the practical aspects of the subjects covered with special emphasis on the site-specific technologies. The curriculum includes courses designed specifically for SRS facilities such as: heat transfer; fluid flow; thermodynamics; electrical science; nuclear physics; nuclear criticality; chemistry; process controls; and facility systems. Some technologies in place at SRS are no longer covered adequately by most of the engineering departments at various universities. The training for newly-assigned engineers is designed in part to address this issue.

The Project Team stated that the Washington Safety Management Solutions (WSMS) personnel assigned to a site facility must receive the same training provided to all other individuals performing similar functions. However, it appears that this training program does not cover the WSMS personnel not assigned to a site facility. It is unclear what training is required and provided for the WSMS personnel not assigned to the site but working on SRS-related projects.

The use of the safety assessment methodologies, together with the thorough knowledge of the facilities being analyzed, will significantly enhance the accuracy and completeness of the Safety Analysis Reports, thus reducing the number of discovery Unresolved Safety Questions. Thus, it is important that WSMS safety analysts receive training as a minimum in: 1) the methodologies (including quantitative human reliability) used in safety assessment; and 2) the facilities being analyzed.

The Project Team acknowledges that the following training issues need to be addressed on a site-wide basis: lack of depth in technical disciplines and/or system specifics; facilities-specific (e.g., process, safety basis) training; continuing training program; lack of training program standardization; and course and/or instructor availability.

The Project Team is commended for its criticality training as documented in the *Westinghouse Savannah River Company Nuclear Criticality Safety Manual* (WSRC 2002b). The Westinghouse Savannah River Company (WSRC) provides its employees with the training and development necessary for them to complete their work assignments in a safe, effective, and quality manner.

Criticality safety training is generally separated into three types of training depending on the employee's duties. The three types of training are: 1) general training; 2) facility-specific training; and 3) operational training. General and facility-specific training are typically based on lectures, films, or video tapes and usually also require written examinations based on the materials presented. In some cases, general and facility-specific training may be combined. Operational training may consist of procedure walkdowns and/or may be included as part of an oral board examination. Employees receiving general and facility-specific criticality safety training also receive Nuclear Incident Monitor (NIM) evacuation alarm familiarization/response training as part of their overall criticality safety training and may be required to participate in NIM evacuation drills.

Whenever the requirements of the *Westinghouse Savannah River Company Nuclear Criticality Safety Manual* (WSRC 2002b) are applicable, each division develops its own criticality safety training programs for its operating facility personnel. Criticality safety training is provided to facility operators; supervisors; fissionable material handlers; accountable material handlers; maintenance mechanics; engineers providing technical or process support; fissionable materials custodians; construction; facility management; lab technicians; and all other personnel whose job assignments require them to be in an area where controls are instituted to ensure criticality safety. Divisions may work together to develop criticality safety training programs and/or train personnel. In some instances, training may be tailored to address only those items required to maintain incredibility of criticality incidents. This is applicable to those facilities in which: 1) all criticality accident scenarios have been judged incredible (such that criticality safety controls are not required); and 2) process conditions and assumptions important to the incredibility of the criticality incidents are to be maintained.

The type, depth, and emphasis of criticality safety training is dependent upon the employee's duties and position. Consideration is given to the nature of a person's assignment. Persons whose responsibilities are managerial are provided training which includes information on how to manage criticality safety concerns. Persons whose responsibilities are technical or operational are provided training which includes information related to the development of technical and operational competence. The training of supervisors is of increased depth as compared to operators/ fissile material handlers. Each division documents the criticality safety qualifications of each job category (education, experience, training, and examination) of staff. Qualifications of such persons are valid for a maximum of two years (unless revoked for cause), at which time personnel shall be requalified.

Criterion 4

Does the SRS Functional Classification process properly classify equipment?

Finding 4 of the RP

The application of the SRS Functional Classification process properly classifies the equipment. This approach is generally conservative resulting in the possible overclassification of some systems. The Project Team reported that to date Functional Classification has been implemented in a conservative manner. Reductions in Functional Classification (provided they are appropriate) can result in significant savings without incurring undue risks.

The Functional Classification process is described in Procedure 2.25 of *Conduct of Engineering and Technical Support* (WSRC 2002h) which provides the: requirements; responsibilities; and methodology for identifying Safety Class (SC) and Safety Significant (SS) functions and the Structures, Systems, and Components (SSC) or Administrative Controls which are credited with performing these functions that reduce the risk to the public and workers from uncontrolled releases of radioactive and other hazardous materials. In addition, it provides methodology for assigning Interim, Preliminary, and Final Functional Classifications to SSCs. It applies to the classification of SSCs in existing facilities, new facilities, and modifications to existing facilities.

This procedure provides the methodology for the selection of non-SC/SS Defense-in-Depth items. Selection of an SSC as non-SC/SS Defense-in-Depth does not alter its classification. The provisions of this procedure apply to WSRC and other members of the Performing Entity (i.e., Westinghouse Savannah River Company, Bechtel Savannah River, Inc., BWXT Savannah River Company, and BNFL Savannah River Corporation) for management and operations at SRS, and to subcontractors performing work for any member of the Performing Entity when required by contract or applicable law.

The Functional Classification of an SSC is determined by this procedure. All SSCs are assigned a Functional Classification of General Services unless it has been determined and documented that another Functional Classification is required.

During the Functional Classification process, SSCs can be classified with varying levels of detail. For instance, when the classification process is initially performed for a facility and a system is classified as SC or SS, the process may stop at that level and all components would be assigned the classification of the system. The process may continue to classify the supporting systems and components separately. The process of classifying the supporting systems and components may be performed over a longer period with each item individually classified only as needed (e.g., a component in a system may not be initially classified individually, but would be classified if there is a need to "touch" the item in the maintenance or replacement arena). This process also applies to components and their sub-components. The classification information is approved by the authoring organization manager and the Design Authority.

Criterion 5

Are the Nuclear Safety Program Performance Indicators meaningful and sufficiently comprehensive?

Finding 5 of the RP

The Nuclear Safety Program Performance Indicators are meaningful to the SRS management for identifying the number of significant nuclear safety issues and the corresponding status of backlog resolution. However, the Nuclear Safety Program Performance Indicators are not sufficiently comprehensive to reflect the SRS nuclear safety culture.

The Nuclear Safety Program Performance Indicators need to be normalized to account for the major facilities that may be shut down during a given time period. By not taking into account the system or facility operating status during the measurement period, the Nuclear Safety Performance Indicators cannot be accurately used for determining the safety posture or trends.

The Nuclear Safety Program Performance Indicators would be of more value if they were expanded to include "minor events" in addition to deviations from nuclear safety control. Recent inclusions of the status and age of nuclear safety issues to the Nuclear Safety Program Performance Indicators have driven many long-term issues to closure and greatly reduced the number of open items. In addition to the Nuclear Safety Program Performance Indicators, there are many other performance indicators tracked by the Project Team that could be used to evaluate nuclear safety performance.

The desktop instructions for Nuclear Safety Metrics are utilized as Nuclear Safety Program Performance Indicators. Currently, they do not reflect the full range of potential indicators that can be used for assessing the Nuclear Safety Program performance.

The new indicator for engineer field time has improved engineering knowledge of the facilities and working relationship with the operations staff.

Criterion 6

Are the Nuclear Safety Program Performance Indicators sufficiently predictive of nuclear safety problems?

Finding 6 of the RP

The Nuclear Safety Program Performance Indicators are not sufficiently predictive of nuclear safety problems. The Nuclear Safety Program Performance Indicators orally presented to the RP do not adequately take into account the potential cumulative effect of "minor incidents" (including radiological and industrial safety events).

The Project Team reported that its system of threshold reporting with regard to criticality occurrences was recently revised to exclude "minor incidents". Repeated occurrence of "minor incidents" that are neither tracked and trended, nor adequately addressed, can result in failure to recognize the possibility that they could be precursors to significant nuclear safety incidents. For example, small breaches (such as leaks in a vessel containing fissile material) coupled with inappropriate operator responses could lead to a nuclear criticality or radiological event.

Criterion 7

Are the threshold values—for entry into the New Information and Potential Inadequacy in Safety Analysis declaration—used in the Unreviewed Safety Question (USQ) Process adequately defined?

Finding 7 of the RP

The Project Team reported during its oral presentation that there are no numerical threshold values used for entry into the New Information and Potential Inadequacy in Safety Analysis declaration as related to the USQ Process. The decision is "knowledge"-based, and therefore the USQ Process involves a significant level of subjectiveness by the affected facility Regulatory Program Manager.

Depending on the complexity of a situation, a considerable time lapse can exist between the initial receipt of New Information and the final determination of whether a Discovery USQ exists. The intent of 10 CFR 830.203 (DOE 2001a) is to avoid excessive time between the receipt of New Information and notification to the DOE. On the other hand, notifying the DOE and taking actions based on unconfirmed information is not desirable. In view of the above, the term "Potential Inadequacy in Safety Analysis (PISA)" is designated as an intermediate point in the total information evaluation process that precedes the declaration of a Discovery USQ, but for which sufficient "potential" exists for a USQ to warrant taking certain precautionary actions. Before a PISA is declared, a reasonable time period is permitted to confirm the reliability of New Information and assess the potential for a Discovery USQ.

Identifying the point in the process at which a PISA is designated cannot be precisely defined; rather, it is based on the judgment of facility management, considering all the circumstances and factors involved. Therefore, it is not necessary to rely on numerical threshold values for this process at this stage of development. The criteria outlined in Procedure 1.04 (Potential Inadequacy in Safety Analysis Declaration Process) and Procedure 1.05 (Nuclear Facility Unreviewed Safety Questions) of the *Facility Safety Document Manual* (WSRC 2001) are adequate.

Criterion 8

Is the USQ Process adequately communicated to the field engineers?

Finding 8 of the RP

The Project Team stated that there is no specific measurement of the adequacy of the communication of the USQ Process to field engineers. However, evidence of adequacy can be inferred from the results of assessments performed by the Quality Assurance Department and the Facility Evaluation Boards.

The results of the assessments performed during the last year have not indicated any deficiency in the understanding of the USQ Process. Assessments are performed in accordance with the *WSRC Source and Compliance Document WSRC-SCD-4* (WSRC 2002g). Assessments are performed periodically (approximately every 18 months) and include review of the 23 functional areas. One of these functional areas, which is assessed by an experienced senior engineer, is Safety Documentation. Several of the criteria to be fulfilled are specifically designed to ensure that the USQ Process is understood. These criteria include the following:

1) A system (i.e., procedures) established for conducting USQ evaluations, review, documentation approval; and control of subsequent changes in safety documents, procedures, or design features; 2) Training of the facility personnel responsible for conducting USQ evaluation/ determination; and 3) USQ determinations conducted, documented, and approved following established procedures.

In addition to the site evaluation process, individual facilities are required to perform comprehensive self-assessments on a continuing basis using the same *WSRC Source and Compliance Document WSRC-SCD-4* (WSRC 2002g) Criteria. Facility self-assessments of the USQ Process have not identified any recurrent deficiencies regarding the understanding or implementation of the USQ Process at a program level.

Moreover, a computerized USQ Process desktop is under development. Utilization of this desktop could enhance the adequacy of communication of the USQ Process to the field engineers.

RECOMMENDATIONS

Based on a careful assessment of the information presented to the Review Panel (RP) and the findings developed in response to the review criteria, the RP provides the following recommendations:

1. The Project Team should develop a single document that identifies all the elements of the Nuclear Safety Program.
2. In the near future, the Project Team should evaluate the effectiveness of the new organization that consolidates all nuclear safety management responsibilities.
3. The Project Team should complete the Safety Analysis Documentation Improvement Strategy (SADIS) in a timely manner.
4. The personnel performance evaluation system should be reviewed to reflect the alignment of nuclear safety responsibilities to individual accountability.
5. The Project Team should review the program for Washington Safety Management Solutions (WSMS) safety analysts assigned to Savannah

River Site tasks to ensure, as a minimum, training in: the methodologies (including quantitative human reliability) used in safety assessment; and the facilities being analyzed.

6. The Project Team should address in a timely manner the following training issues on a site-wide basis: lack of depth in technical disciplines and/or system specifics; facilities-specific (e.g., process, safety basis) training; continuing training program; lack of training program standardization; and course and/or instructor availability.
7. The Nuclear Safety Program Performance Indicators should be normalized to account for the major facilities status during the measurement period.
8. The Nuclear Safety Metrics should reflect the full range of potential indicators that can be used for assessing the Nuclear Safety Program performance.
9. The Project Team should consider accounting for the potential cumulative effect of "minor incidents" (including radiological and industrial safety events) into the Nuclear Safety Program Performance Indicators.
10. The Project Team should complete in a timely manner its computerized Unreviewed Safety Question (USQ) Process desktop.

References

ANS (American Nuclear Society). Nuclear criticality control of special actinide elements, ANSI/ANS-8.15-1981. La Grange Park, IL: ANS; 1981.

ANS (American Nuclear Society). Safety in conducting subcritical neutron multiplication measurements in situ, ANSI/ANS-8.6-1983. La Grange Park, IL: ANS; 1983.

ANS (American Nuclear Society). Criticality safety criteria for the handling, storage, and transportation of LWR fuel outside reactors, ANSI/ANS-8.17-1984. La Grange Park, IL: ANS; 1984.

ANS (American Nuclear Society). Nuclear criticality safety training, ANSI/ANS-8.20-1991. La Grange Park, IL: ANS; 1991.

ANS (American Nuclear Society). Use of fixed neutron absorbers in nuclear facilities outside reactors, ANSI/ANS-8.21-1995. La Grange Park, IL: ANS; 1995.

ANS (American Nuclear Society). Use of borosilicate glass raschig rings as a neutron absorber in solutions of fissile material, ANSI/ANS-8.5-1996. La Grange Park, IL: ANS; 1996a.

ANS (American Nuclear Society). Administrative practices for nuclear criticality safety, ANSI/ANS-8.19-1996. La Grange Park, IL: ANS; 1996b.

ANS (American Nuclear Society). Criticality accident alarm system, ANSI/ANS-8.3-1997. La Grange Park, IL: ANS; 1997a.

ANS (American Nuclear Society). Nuclear criticality safety based on limiting and controlling moderators, ANSI/ANS-8.22-1997. La Grange Park, IL: ANS; 1997b.

ANS (American Nuclear Society). Nuclear criticality safety in operations with fissionable materials outside reactors, ANSI/ANS-8.1-1998. La Grange Park, IL: ANS; 1998a.

ANS (American Nuclear Society). Guide for nuclear criticality safety in the storage of fissile materials, ANSI/ANS-8.7-1998. La Grange Park, IL: ANS; 1998b.

ASME (The American Society of Mechanical Engineers). Assessment of technologies supported by the U.S. Department of Energy; Results of the peer review for fiscal year 1997, vol. 47. Washington, DC: Center for Research and Technology Development; New York: ASME; 1997.

ASME (The American Society of Mechanical Engineers). Assessment of technologies supported by the U.S. Department of Energy; Results of the peer review for fiscal year 1998, vol. 50. Washington, DC: Center for Research and Technology Development; New York: ASME; 1998.

ASME (The American Society of Mechanical Engineers). Assessment of technologies supported by the U.S. Department of Energy; Results of the peer review for fiscal year 1999, vol. 56. Washington, DC: Center for Research and Technology Development; New York: ASME; 1999.

ASME (The American Society of Mechanical Engineers). Manual for peer review. Washington, DC: ASME; 2000a.

ASME (The American Society of Mechanical Engineers). Assessment of technologies supported by the U.S. Department of Energy; Results of the peer review for fiscal year 2000, vol. 61. Washington, DC: Center for Research and Technology Development; New York: ASME; 2000b.

ASME (The American Society of Mechanical Engineers). Assessment of technologies supported by the U.S. Department of Energy; Results of the peer review for fiscal year 2001, vol. 64. Washington, DC: Center for Research and Technology Development; New York: ASME; 2001a.

ASME (The American Society of Mechanical Engineers). Strategy for remediation of groundwater contamination at the Nevada Test Site, CRTD-Vol. 62. New York, NY: ASME; 2001b.

ASME (The American Society of Mechanical Engineers). Requirements for disposal of remote-handled transuranic wastes at the Waste Isolation Pilot Plant, CRTD-Vol. 63. New York, NY: ASME; 2001c.

ASME (The American Society of Mechanical Engineers). Waste Isolation Pilot Plant initial report for polychlorinated biphenyl disposal authorization, CRTD-Vol. 65. New York, NY: ASME; 2002a.

ASME (The American Society of Mechanical Engineers). Airborne release fractions, CRTD-Vol. 68. New York, NY: ASME; 2002b.

ASME (The American Society of Mechanical Engineers). The beryllium oxide manufacturing process, CRTD-Vol. 69. New York, NY: ASME; 2002c.

DOE (U.S. Department of Energy). DOE Order 5480.23. Washington, DC: U.S. Department of Energy; 1991.

DOE (U.S. Department of Energy). Planning and preparedness for operational emergencies, DOE Order 5500.3A. Washington, DC: DOE; (February) 1992.

DOE (U.S. Department of Energy). Guidance for preparation of DOE 5480.22 and DOE 5480.23 implementation plans, DOE-STD-3011-94. Washington, DC: DOE; 1994.

DOE (U.S. Department of Energy). Hazard categorization and accident analysis techniques for compliance with DOE order 5480.23, nuclear safety analysis reports, DOE-STD-1027-92, change notice no. 1. Washington, DC: DOE; (September) 1997.

DOE (U.S. Department of Energy). Guidelines for preparing criticality safety evaluations at DOE non-reactor nuclear facilities, DOE-STD-3007-93. Washington, DOE; (November) 1998.

DOE (U.S. Department of Energy). Preparation guide for U.S. Department of Energy nonreactor nuclear facility safety analysis reports, DOE-STD-3009-94. Washington, DC: DOE; (January) 2000a.

DOE (U.S. Department of Energy). Nuclear criticality safety, DOE Order 420.1. Washington, DC: U.S. Department of Energy; (November 22) 2000b.

DOE (U.S. Department of Energy). Program and project management for the acquisition of capital assets, DOE Order 413.3. Washington, DC: DOE; (October 13) 2000c.

DOE (U.S. Department of Energy). Nuclear safety management, 10 CFR 830. Washington, DC: DOE; (February 9) 2001a.

DOE (U.S. Department of Energy). Personnel selection, qualification, training, and staffing requirements at DOE reactor and non-reactor nuclear facilities, DOE Order 5480.20A. Washington, DC: DOE; (July 12) 2001b.

EPA (U.S. Environmental Agency). Designation, reportable quantities, and notification, 40 CFR 302. Washington, DC: U.S. Government Printing Office; (July 1) 1999.

EPA (U.S. Environmental Agency). Chemical accident prevention provisions, 40 CFR 68. Washington, DC: U.S. Government Printing Office; (January1) 2002.

OSHA (U.S. Occupational Safety and Health Administration). Hazardous waste operations and emergency response and health and safety for hazardous waste operations, 29 CFR 1910.120. Washington, DC: U.S. Government Printing Office, 2002a.

OSHA (U.S. Occupational Safety and Health Administration). Process safety management (PSM) of highly hazardous chemicals, 29 CFR 1910.119. Washington, DC: U.S. Government Printing Office 2002b.

RSI (Institute for Regulatory Science). Handbook of peer review. Columbia, MD: RSI; 2000.

Savannah River Operations Office. Inside Savannah River, July 2, 2002. Available at http://sro.srs.gov/inside1.htm

SRS (Savannah River Site). Historical highlights @ SRS, April 23, 2002. Available at http://www.srs.gov/general/about/hilits.htm

USNRC (U.S. Nuclear Regulatory Commission). Packaging and transportation of radioactive material, 10 CFR 71. Washington, DC: U.S. Government Printing Office; 2002.

WSMS (Westinghouse Safety Management Solutions). WSMS nuclear criticality safety methods manual, WSMS-CRT-01-0116. Rev.0. Aiken, SC: Westinghouse Savannah River Company; (October) 2001.

WSRC (Westinghouse Savannah River Company). Facility safety document manual, WSRC-11Q. Aiken, SC: Westinghouse Savannah River Company; 2001.

WSRC (Westinghouse Savannah River Company). Nuclear and process safety manual, WSRC-1-01. Aiken, SC: Westinghouse Savannah River Company; 2002a.

WSRC (Westinghouse Savannah River Company). WSRC nuclear criticality safety manual. WSRC-SCD-3. Aiken, SC: Westinghouse Savannah River Company; 2002b.

WSRC (Westinghouse Savannah River Company). Standard/requirements identification document (S/RID), WSRC-RP-94-1286. Aiken, SC: Westinghouse Savannah River Company; 2002c.

WSRC (Westinghouse Savannah River Company). Standard for development and maintenance of hazard assessment, WSRC EMPP 6Q-001. Aiken, SC: Westinghouse Savannah River Company; (July) 2002d.

WSRC (Westinghouse Savannah River Company). SRS process safety management manual, WSRC-IM-90-135. Aiken, SC: Westinghouse Savannah River Company; (November) 2002e.

WSRC (Westinghouse Savannah River Company). Transportation safety manual, WSRC-19Q. Aiken, SC: Westinghouse Savannah River Company; (September) 2002f.

WSRC (Westinghouse Savannah River Company). WSRC source and compliance document, WSRC-SCD-4. Aiken, SC: Westinghouse Savannah River Company; 2002g.

WSRC (Westinghouse Savannah River Company). Conduct of Engineering and Technical Support. WSRC-E7. Revision 11. Aiken, SC: Westinghouse Savannah River Company; (April 1) 2002h.

WSRC (Westinghouse Savannah River Company). Industrial hygiene manual, WSRC-4Q. Aiken, SC: Westinghouse Savannah River Company; (January) 2003.

Acronyms and Definitions

AA: Authorization Agreement
The AA, initiated by the U.S. Department of Energy (DOE)-Screening Reviewer (SR), establishes and documents in one concise document the terms and conditions that are binding on Westinghouse Savannah River Company for a particular scope of operation at the selected facilities in a manner that preserves DOE's bases for authorizing facility operations.

AC: Administrative Control
AC is a means of control which relies on the judgement, training, and responsibility of people for implementation. Administrative means of control may be action, caution, or verification steps in a procedure, or steps in a surveillance program.

AL: Administrative Limit
ALs constitute a graded approach to providing operating limits for non-nuclear facilities or other industrial facilities.

ASA: Auditable Safety Analysis
ASA documentation is a graded approach to providing safety analysis documents for chemical and radiological facilities.

ASME: American Society of Mechanical Engineers

BIO: Basis for Interim Operation
BIO can represent the Documented Safety Analysis for short-life operations or facilities in transition from operating missions.

BNFL: British Nuclear Fuels, Limited Savannah River Corporation

BSRI: Bechtel Savannah River, Inc.

BWXT: BWX Technologies, Inc.

B&WSRC: B&W Savannah River Company

CCB: Configuration Control Board

CSCs: Criticality Safety Committees
CSCs are maintained in the Nuclear Materials Management Division, High Level Waste Division, and Savannah River Technology Center to: assist in self-assessment; provide technical support; and perform in-process review as necessary regarding fissionable material handling, storage, processing, and disposal activities from a criticality safety perspective.

DATR: Design Authority Technical Review

DBA: Design Basis Accident

DCOP: Disciplined Conduct of Projects

DID: Defense-in-Depth

DOE: U.S. Department of Energy

DSA: Documented Safety Analysis
DSA is a documented analysis of the extent to which a HC-1, HC-2, HC-3 Nuclear Facility can be operated safely with respect to workers, the public, and the environment, including a description of the conditions, safe boundaries, and hazard controls that provide the basis for ensuring safety. A DSA normally consists of either a Safety Analysis Report or a Basis for Interim Operation.

EMP: Emergency Management Program

EO: Evaluation Originator
EO is an individual who meets the appropriate qualification and training requirements and is assigned technical responsibility by facility management to perform Unreviewed Safety Question Evaluations (USQEs).

EP: Executive Panel

EPHA: Emergency Preparedness Hazard Assessments
EPHAs document a special hazard assessment that satisfies emergency management program (EMP) requirements. This assessment is used to determine the extent and scope of emergency planning and preparedness activities.

ESH&QA: Environment, Safety, Health, and Quality Assurance

FDD: Facilities Disposition Division

FEB: Facility Evaluation Boards

FHA: Fire Hazard Analysis
FHA documents a special fire hazard assessment and is required for nuclear facilities, all new facilities, and any major modification of existing facilities.

FHC: Facility Hazard Category
FHC is a hazard measure based primarily on the quantities, locations, and forms of radiological and hazardous material inventories present. Depending upon the level of hazard, the hazard category is designated as HC-1, HC-2, or HC-3.

FOSC: Facilities Operations Safety Committees
FOSCs are responsible for reviewing and approving all Unreviewed Safety Question Evaluations (USQEs).

GS: General Services

GSAR: Generic Safety Analysis Report

HAD: Hazard Assessment Document
HAD presents the results of hazard identification and categorization and includes the inventory and Facility Hazard Category (FHC) along with associated facility segmentation.

HASP: Health and Safety Plan
HASPs are required by Occupational Safety and Health Administration (OSHA) regulation. A site-specific HASP can provide essential hazard analysis information so that such information need not be repeated in other documents.

HC: Hazard Category (See FHC)

HLWD: High Level Waste Division

IME: Integrated Management Evaluation

ISMS: Integrated Safety Management System

ITS: Important to Safety

JCO: Justification for Continued Operation
JCOs are used both to document the current safety envelope for facilities that do not conform to current safety documentation requirements, and to provide an interim basis if an extended analysis is needed during the Unreviewed Safety Question (USQ) process.

LCO: Limiting Condition for Operation

LD: Linking Document

ME: Management Evaluation

MP: Management Policy

NCSE: Nuclear Criticality Safety Evaluation

NI: New Information
NI is information that might jeopardize the adequacy of the Self-Assessment that supports the Safety Basis for a nuclear facility.

NIM: Nuclear Incident Monitor

NMMD: Nuclear Materials Management Division

NSDS: Nuclear Safety Data Sheet

Nuclear Criticality Safety Review Committee
NCSRC, reporting to the Technical and Engineering Management Integration Council, is the highest level site criticality safety committee and is responsible for fostering criticality safety across the site and serving as an arbiter in resolving criticality safety technical and policy issues.

ORR: Operational Readiness Review
ORRs for nuclear facilities are performed prior to initial startup or prior to restart following shutdown and major modifications.

OSR: Operational Safety Requirement

PA: Proposed Activity

PE&C: Projects Engineering & Construction

PE&CD: Projects Engineering and Construction Division

PHA: Preliminary Hazard Analysis
PHA is developed as support for HC-1, HC-2, HC-3 Nuclear Facilities Safety Analysis Reports or Basis for Interim Operations (BIOs).

PHR: Process Hazard Review
PHR predates the OSHA-Process Safety Management or U.S. Environmental Protection Agency-Risk Management Program rules, and it is conducted at the discretion of the facility on many operations regardless of chemical inventories.

PISA: Potential Inadequacy in Safety Analysis
PISA is designated as an intermediate point in the total information evaluation process that precedes the declaration of a Unreviewed Safety Question (USQ), but, for which sufficient "potential" exists for a USQ to warrant taking certain precautionary actions.

POS: Plant Operating Specification

PRCEE: ASME Peer Review Committee for Energy and the Environment

PrHA: Process Hazard Analysis
PrHA is required for any processes involving hazardous chemicals at or above the specified Threshold Quantities.

PS: Production Support

PSM: Process Safety Management

QA: Quality Assurance

QR: Qualified Reviewer
QR is an individual assigned by facility management who meets the qualifications and training requirements.

RMP: Risk Management Program
RMP designates a program required by EPA regulation 40 CFR 68, *Accidental Release Prevention Provisions.*

RMPlan: Risk Management Plan
RMPlan includes a summary of Process Hazard Analysis (PrHA) as required by the Risk Management Program (RMP).

RP: Review Panel

ROSC: Reactor Operation Safety Committee

RQ: Reportable Quantity

RSI: Institute for Regulatory Science

SA: Self-Assessment

SADIS: Safety Analysis Documentation Improvement Strategy

SAR: Safety Analysis Report
SAR is the primary Safety Basis document for a nuclear facility and presents the results of hazard evaluation and accident analyses, and supporting information to demonstrate that the facility as designed and constructed can be operated without undue risk to workers, the public, and the environment.

SARP: Safety Analysis Report for Packaging
SARP is a special Safety Analysis Report (SAR) that addresses transportation.

SB: Safety Basis
The compilation of essentially all safety-related documents used to design, implement, and evaluate safety aspects at a nuclear facility.

SC: Safety Class

SER: Safety Evaluation Report

SO: Screening Originator
SO is an individual who is assigned technical responsibility by facility management for performing the Unreviewed Safety Question Screening (USQS) and initiating changes to Safety Basis (SB) documents that are identified as a result of the Unreviewed Safety Question (USQ) process.

SR: Screening Reviewer
SR is an individual who is assigned technical responsibility for performing the Unreviewed Safety Question Screening (USQS) Reviews.

S/RIDs: Standards/Requirements Identification Documents

SRS: Savannah River Site

SRTC: Savannah River Technology Center

SS: Safety Significant

SSC: Structures, Systems, and Components

SWD: Solid Waste Division

TEMIC: Technical and Engineering Management Integration Council

TQ: Threshold Quantities

TSR: Technical Safety Requirement
TSRs replace Operational Safety Requirement (OSRs), their bases, and the management or Administrative Controls (ACs) required to assure safe operation. Nuclear Facilities are operated in accordance with TSRs.

TSRs for HC-3 facilities may consist only of inventory control and commitments to safety programs.

USQ: Unreviewed Safety Question
USQ is a key safety issue consisting of a situation that has not been previously reviewed or approved by DOE, and meets one or more of the conditions identified in DOE regulations 10 CFR 830.

USQE: USQ Evaluation
USQE involves a series of steps and approval actions and may result in a determination that a Unreviewed Safety Question (USQ) exists.

USQS: USQ Screening
USQS is used to identify those proposed activities that require a Unreviewed Safety Question Evaluation (USQE).

WSMS: Washington Safety Management Solutions

WSRC: Westinghouse Savannah River Company

Biographical Summaries

Gary A. Benda* is President of U.S. Energy Corp.—an environmental management firm specializing in radioactive mixed waste management, health physics, decontamination and decommissioning, and technology development. Previously, he was Vice-President, General Manager of the Programs Division for NUKEM Nuclear Technologies, Inc. His responsibilities included developing and maintaining federal programs in North America that specialized in engineering and waste-processing services. Prior to NUKEM, he spent over 17 years with Chem-Nuclear Systems/WMX Technologies in various management roles. He also directed the site investigation, geophysical analysis, site screening, and license application, as well as managed the public hearings and licensing operations associated with local and national regulatory agencies for new low-level waste sites. He has over 20 years of experience in environmental restoration, technology development, and waste management, and has instructed over 20 national and international professional courses on radioactive waste management, mixed waste, and technology development. He is a member of the American Society of Mechanical Engineers (ASME), American Nuclear Society, and Health Physics Society. He has served as Chair of the ASME National Mixed Waste Committee, Environmental Remediation Committee, and Environmental Engineering Division. He has also chaired over 100 technical sessions at numerous national and international conferences on environmental management. He has authored and coauthored various scientific papers, reports, book chapters, and articles on the nuclear environment. Gary Benda is a Certified Health Physicist. He received a B.S. in Health Physics from Oklahoma State University, an M.S. in Applied Nuclear Science from Georgia Institute of Technology, and an M.B.A. from Seattle City University.

Erich W. Bretthauer* is currently President of the Bryce Meadows Development Corporation. He held the position of research professor at the University of Nevada-Las Vegas from January 1993 to March 1995. In that capacity, he served as Executive Director of Nevada Industry, Science Engineering & Technology, a public-private partnership which developed programs to enhance the scientific infrastructure of the state of Nevada. He was Assistant Administrator for Research and Development at the U.S. Environmental Protection Agency (EPA) from March 1990 until January 1993. In that capacity, he managed the Research and Development activities of a large and multi-disciplinary agency. Erich

Bretthauer rose through the ranks of the EPA and served in a number of capacities ranging from a bench scientist to policy manager at national and international levels. He directed the EPA's emergency and long-term monitoring program after the accident at Three Mile Island, as well as its bioremediation program in Prince William Sound after the Valdez oil spill. He also directed the EPA's ecological research program from 1983-1986 and was Director of EPA's Environmental Monitoring Systems Laboratory in Las Vegas from 1986-1990. He is a member of Sigma Xi; the American Chemical Society; the American Association for the Advancement of Science; and the American Water Works Association; and has served on the Federal Advisory Committee to the Civil Engineering Research Foundation. Erich Bretthauer is the author and coauthor of numerous papers, reports, and other publications. He received his B.S. and M.S. in chemistry from the University of Nevada, Reno, NV.

Ernest L. Daman* is Chairman Emeritus of Foster Wheeler Development Corporation where he previously served as Director of Research and Chairman of the Board. He also held the position of Senior Vice President at the parent company, FWC. He is a Past President of the American Society of Mechanical Engineers and was elected to the National Academy of Engineering. Ernest Daman is a Fellow of the Institute of Energy (England) and the American Association for the Advancement of Science, and Past Chairman of the American Association of Engineering Societies. He served on several American Society of Mechanical Engineers committees as member or chairman. Ernest Daman is the author of numerous papers and holds 18 patents. He was responsible for the design and development of a combined steam gas turbine plant; fluidized bed combustion; fast breeder reactor components; supercritical steam generators; environmental control processes; and advanced high-efficiency power generation systems. Ernest Daman received his B.M.E. degree from the Polytechnic Institute of Brooklyn.

James W. Dickey is a consultant with expertise in transportation; the storage and processing of spent reactor fuel; and environmental compliance management. He has evaluated a plutonium facility for restart after a shutdown for safety concerns; provided engineering and design support for the decontamination and removal of contaminated ventilation systems for a nuclear fuel processing facility; as well as provided guidance and reviewed plans for compliance requirements with the U.S. Department of Energy and the U.S. Environmental Protection Agency for the decommissioning of a research reactor. James Dickey has been a consultant to the U.S. Department of Energy, coordinating data acquisition and preparing the first Baseline Environmental Management Report to Congress—which addressed the cost and scheduling for clean-up of the Savannah River Site; System Engineering Manager for high-level waste at the Savannah River Site, providing expert advice on all technical and safety aspects associated with high-level waste engineering systems; project management on two major Savannah River Site projects involved in the separation of uranium isotopes using advanced technology, and uranium oxide fuel fabrication; Technical Authority on U.S. Department of Energy Orders 6430.1A, "General Design Criteria"; "Life-Cycle Asset Management" for high-level waste facilities and projects, providing coordination and technical direction to division staff in the oversight of contractor engineering activities; and Branch Chief for the Liquid Waste Division, as program/project manager for radioactive liquid waste storage, transfer, treatment, and disposal. James Dickey also served as a supervisor in Nuclear Field Engineering and as an Engineer and Supervisory Engineer in the Nuclear Fluid Systems and Mechanical Division. His various job responsibilities have included: evaluation and approval of safety analysis reports; processing hazards analysis; engineering calculations; start-up test plans; test results; and special procedures. He was also responsible for the operation of high-level waste tank farms; evaporators; an in-tank precipitation facility; an extended sludge processing facility; an effluent treatment facility; a new waste transfer facility; waste removal from storage tanks; replacement of high-level waste evaporators; and various facility upgrade projects, such as upgrading waste tank ventilation systems to the latest nuclear criteria. James Dickey was the Deputy Director of the U.S. Department of Energy's New Production Reactor Project Management Office, where he participated/directed/coordinated activities in all aspects of the New Production Reactor project. He was also connected with the Project Engineering Division of Savannah River Site, where he was involved in projects that included new-facility site selection based on geotechnical data (cone penetrometer; soil characterization—including dense nonaqueous phase liquids; groundwater elevation/flow; and surface and subsurface radioactive contamination). His position involved the development of environmental criteria and permitting strategy; maintenance and repair procedures for nuclear equipment and systems; inspection procedures; acceptance criteria; procurement specifications; management and direction of engineering support for maintenance, operations, testing, and refueling of nuclear submarines; technical input to the Environmental Impact Statement; preparation of the draft project plan and project management plan; construction or modification of facilities related to nuclear reactors; weapons material production; fuel fabrication; power production; waste treatment; environment compliance; and infrastructure. James Dickey also certified the official readiness of all nuclear fluid systems and mechanical equipment for start-up testing and operation of a reactor plant; participated in seven reactor plant start-ups after 100% refueling, and 11 reactor plant start-ups after extended maintenance and overhaul outages; and reviewed and approved shipyard facility design and modifications for facilities in which nuclear work was accomplished. He also designed nuclear support systems and developed operating and test procedures; designed and constructed reactor plant fluid system decontamination systems and equipment; and prepared operating procedures. James Dickey studied Chemical Engineering at both undergraduate and graduate levels at Texas A&M University, Kingsville, TX, and received a Ph.D. in Engineering Management from Hamilton University in Evanston, WY. He is a registered Professional Engineer.

William S. Durant is currently a consultant to the Angel Medial Center in Franklin, North Carolina. His current duties include reviewing the Safety Manual to determine potential deficiencies; and designing and fabricating several pieces of equipment for use in the Occupational Therapy Department. Previously, he spent his entire professional career working at the Savannah River Laboratory for two succeeding contractors who include E.I. duPont and the Westinghouse Savannah River Company respectively. With Westinghouse, he held the position of Senior Advisory Engineer. He was the appointed chairman of the risk assessment methodology group for all Westinghouse DOE M&O sites; the appointed chairman of the data banks for risk assessment subcommittee for the DOE Energy Facility Contractors Group; and the appointed Division Technical Advisor whose goals were to examine new processes for technical and safety inadequacies to prevent expensive retrofits, and to mentor less experienced personnel. He served as contractor representative to author DOE-STD-3009 for Safety Analysis Report format and content; served as contractor representative to author SAFT-0029 for hazards categorization standard for EM facilities; and served as technical lead in developing the Integrated Safety Analysis manual for the U.S. Nuclear Regulatory Commission. With E.I. duPont, he conceived and maintained the non-reactor fault tree data banks and sorting codes. The banks contain >400,000 accidents, equipment failures, and process upsets. He coordinated the development and use of Systems Analyses and Safety Analysis Reports for the non-reactor facilities at the Savannah River Plant. Additionally, he developed methodology for the Nuclear Regulatory Commission on risk assessment for commercial reprocessing facilities. He co-authored the Systems Analyses for the primary fuel reprocessing facilities; performed risk assessments for the uranium calcination and tritium facilities; developed methodology for the Nuclear Regulatory Commission on risk assessment for commercial reprocessing facilities; and performed numerous risk-benefit analyses on non-reactor facilities resulting in significant reductions in accident potential. He has also consulted the Parsons Savannah Services Company and the General Physics Corporation on data bank development for risk assessment; and consulted Radian International on a safety review of commercial nuclear fuel reprocessing facility design. William Durant served as Lieutenant in the U.S. Army Corps of Engineers as an Inorganic Chemist, Office of the Chief of Engineers. He was a former member of AIChE professional society, Phi Lambda Upsilon academic society, and the Sigma Xi research society. He served as a consultant to the Advisory Committee for Reactor Safeguards on reactor core meltdown analysis and on the Reactor Safety Study (WASH-1400). He also served on peer review panels for EPRI non-reactor risk assessment studies, INTERTRAN transportation risk calculation system, Guide to Radiological Accident Considerations for Siting and Design of DOE Nonreactor Facilities, Rocky Flats safety analysis guide. He was the first recipient of the Donald A. Orth Award for continuing technical excellence. William Durant has authored and co-authored over 300 publications, reports, and papers including acting as the lead author for the Savannah River Reactor Safety Analysis Report. His work centers on two major technical areas. Earlier publications are associated with heat transfer and hydraulics of nuclear reactors; with confinement of radioactivity following a nuclear reactor accident; and with the behavior of debris following a reactor core meltdown. Later publications are associated with risk analysis of non-reactor nuclear facilities; probabilistic risk assessment methodology development and application; development of data banks and statistics for risk assessment; and accident progression and human factors. He received both B.S. and M.S. degrees in Chemical Engineering from Auburn University.

Irwin Feller* is Director of the Institute for Policy Research and Evaluation and Professor of Economics at The Pennsylvania State University, where he has been on the faculty since 1963. His current research interests include the economics of academic research, the University's role in technology-based economic development, and the evaluation of federal and state technology programs. He is the author of *Universities and State Governments*: A Study in Policy Analysis, and over 100 refereed journal articles, final research reports, book chapters, reviews, and numerous papers presented to academic, professional, and private organizations. He is former Chair of the Committee on Science, Engineering, and Public Policy, American Association for the Advancement of Science. Irwin Feller was the American Society of Mechanical Engineers Pennsylvania State Fellow in 1996-1997. He has been appointed to the National Research Council's Committee on Science, Engineering, and Public Policy; International Benchmarking of U.S. International Competitiveness-Immunology; Transportation Research Board, Research and Technology Coordinating Committee, National Research Council; and National Institute of Standards and Technology-Manufacturing Extension Partnership National Advisory Board. Irwin Feller is Chair of the National Science Foundation's Advisory Committee on Social, Behavioral, and Economic Sciences. He received a B.B.A. in Economics from the City University of New York and a Ph.D. in Economics from the University of Minnesota.

Robert A. Fjeld* is Dempsey Professor of Environmental Engineering and Science at Clemson University. He coordinates the Department's nuclear environmental focus area, which is concerned with the environmental aspects of nuclear technologies including health physics, radioactive waste management, and risk assessment. Previously, he served as a faculty member in the Nuclear Engineering Department at Texas A&M University. He has active research on actinide transport in soils, instrumentation for measuring radioactivity in environmental samples, and environmental risk assessment. Robert Fjeld is a member of the Health Physics Society, American Nuclear Society, Society for Risk Analysis, and the American Society of Mechanical Engineers, where he serves as newsletter editor for the Mixed Waste Committee. He has served on two National Research Council Committees that studied decontamination and decommissioning issues. Robert Fjeld is the author or coauthor of over 80 technical publications and presentations on topics such as radiation measurements, environmental transport of radionuclides, risk assessment, and aerosol physics. Robert Fjeld received a B.S. degree from North Carolina State University; and an M.S. degree and a Ph.D. from The Pennsylvania State University, all in Nuclear Engineering. He is a registered Professional Engineer.

William T. Gregory, III* is currently Principal of Vinculum Marketing Solutions. Prior to forming Vinculum, he was Director of Government Programs for Foster Wheeler Environmental Corporation, an engineering and construction firm providing environmental and waste management services to government and private sector clients worldwide. Previously, he held a number of operational and business development positions at equipment manufacturing and service provision firms supporting nuclear utilities, industrial and process industries, and government agencies. His work has involved the management, processing, and disposition of hazardous, radioactive, and mixed wastes. He has also worked on the decontamination and decommissioning of nuclear facilities and on providing a wide range of environmental services in response to regulatory drivers. Prior to entering the private sector, he served with the U.S. Navy on nuclear submarines and at the operational command center for submarine operations in the Atlantic Fleet. William Gregory is actively involved with a number of international, national, and local organizations including the American Society of Mechanical Engineers and the American Nuclear Society. He is a founding member of the Board of Directors for the annual international Waste Management Symposium. William Gregory has served as an elected officer of several American Society of Mechanical Engineers divisions. He received a B.S. degree in Geology from the University of New Mexico, and an M.B.A. degree from Lamar University. He also attended naval nuclear power, nuclear weapons, and engineering schools as a U.S. Navy officer.

Nathan H. Hurt* is a consultant in management and engineering with Technical and Management Consulting. He provides services to industrial firms and government agencies involved in environmental clean-up and waste management—both chemical and radioactive. He has extensive experience in the areas of executive management; plant management; engineering management; project management; marketing; and sales. He specializes in the areas of: uranium enrichment/production; engineering; development and marketing; plant management of rubber chemicals; petrochemicals; and thermoplastics. He also specializes in the engineering management of synthetic rubber and lattices; vinyl monomers and copolymers; polyesters; U.S. Department of Energy (DOE) weapons plants; quality assurance management; and operational readiness review. Nathan Hurt has been involved with the decommissioning of nuclear facilities. He was the Corporate Sponsor or Program Manager for seven decommissioning contracts at the DOE Complexes in Oak Ridge, TN; and Pinellas, FL. Previously, Nathan Hurt worked for Sharp and Associates, Inc. as the Director and Project Manager at the Oak Ridge Office. He was Vice President and Director of Oak Ridge Operations for IDM Environmental Corp., where he was responsible for the marketing and sales of decontamination, decommissioning, and waste management. He served as Project Manager for the laboratory quality assurance program at Westinghouse Hanford; DOE's Rocky Flats Plant—plant-wide identification of electrical equipment. He managed a study for a waste treatment and storage facility at the Portsmouth Area Uranium Enrichment Facility which included incineration and compaction of low-level radioactive wastes. He also worked for The Goodyear Tire and Rubber Company, including Goodyear Atomic, as Director of Research and Development, and President, where he was responsible for the operation of the Portsmouth Area Uranium Enrichment Facility. Nathan Hurt is a Past President of the American Society of Mechanical Engineers. He has been a member of: the American Association of Engineering Societies' Board of Governors; the American Institute of Chemical Engineers; and the Institute of Nuclear Materials Management. He is also a member of Tau Beta Pi Honorary Engineering Society; Pi Tau Sigma Honorary Mechanical Engineering Society; and was a member of The Nuclear Engineering Advisory Board of Worcester Polytechnic Institute. Nathan Hurt received a B.S. degree in Mechanical Engineering from the University of Colorado and has done Graduate, Technical, and Management course work at Pennsylvania State University. He is a registered Professional Engineer in Ohio.

Michael C. (M.C.) Kirkland is Vice President for the Southeastern Region of the Institute for Regulatory Science (RSI). In that capacity he leads various RSI projects related to the RSI mission in the southeastern U.S. Previously he was an independent consultant involved in peer review and various independent studies. For example, he led a team that performed an External Independent Review of the $1.3 billion Spallation Neutron Source Project at Oak Ridge, TN. He assisted in the planning and review of a management assessment at a U.S. Department of Energy (DOE) Site that involved the restart of a plutonium facility. He participated in planning, procurement, and review activities in the environmental remediation area that included decommissioning activities at a shut-down nuclear test reactor; and designed and installed a ground water cleanup technology. M.C. Kirkland managed several environmental and construction projects that employed many soil investigative techniques including significant work with cone penetrometers. Additionally, he provided consulting services to a large environmental remediation services company regarding Dense Non-Aqueous Phase Liquid locating and removal techniques. During his tenure at the Savannah River Site (SRS) of DOE, M.C. Kirkland was a Technical Advisor, Project Manager, and Director of the Project Engineering Division. He evaluated nuclear and mixed waste conditions and aspects of high level wastes and spent nuclear fuel; determined material inventories; performed pollution prevention and environmental health and safety evaluations for a proposed waste treatment facility; served as technical advisor to a study administered by the Savannah River Operations Office; and developed integrated schedules defined for this project. M.C. Kirkland was director of the Project Engineering Division and managed the SRS design and construction program. He has been involved with waste management and environmental projects; cutting-edge technology programs; and worked with lasers and magnetic containment. He served as Director of the Waste and Fuel Cycle Technology Office, and planned and coordinated the programs of the DOE National High Level Waste Technology Office; the SR Fuel Cycle Technology Program; and the Commercial Interim Spent Fuel Management Program. M.C. Kirkland holds a B.S. in Mechanical Engineering from the University of South Carolina. He is registered as a Professional Engineer in South Carolina.

Peter B. Lederman* is a consultant with over 48 years of experience in all facets of process engineering, environmental management, control, and policy development. This includes hazardous substance management; environmental remediation; environmental audit; pollution prevention; development of air pollution control devices; and reuse of waste products. He recently retired as Executive Director of the Center for Environmental Engineering & Science, Executive Director for Patents and Licensing, and Research Professor of Chemical Engineering and Environmental Policy at the New Jersey Institute of Technology. Peter Lederman managed major programs in industrial waste treatment research and development, and in oil and hazardous material spill control and remediation. Most recently, he was responsible for a study of the Economic Impact of Environmental Regulations. He has been responsible for technology transfer efforts including the maturing and licensing of innovative environmental technologies. He is a Fellow of the American Institute of Chemical Engineers; a Diplomat of the American Academy of Environmental Engineers; and a member of the American Society of Mechanical Engineers. He has served on several committees of the NRC and is the chair of the U.S. National Research Council Committee (USNRC) on Review and Evaluation of the Army Chemical Stockpile Disposal Program. He chaired the American Institute of Chemical Engineer's Environmental Division and is currently chair of its Societal Impacts Operating Council. Peter Lederman received a B.S.E., M.S.E., and Ph.D. (all in Chemical Engineering) from the University of Michigan in Ann Arbor, MI and is a registered Professional Engineer.

Betty R. Love is currently Executive Vice President of the Institute for Regulatory Science. In that capacity, she is responsible for the management of day-to-day operations of the Institute, and for administration of several projects. She is the Administrative Manager of a large-scale peer review program in collaboration with the American Society of Mechanical Engineers for a number of organizations including the U.S. Department of Energy. Her current research activities center around the development and implementation of a systematic approach to stakeholder participation, notably in scientific meetings. Previously, Betty Love was Director, Department of Training and Information within the Office of Environmental Health and Safety of Temple University in Philadelphia, PA. During that period she was instrumental in the development of a "Handbook of Environmental Health and Safety". She also developed and implemented a large-scale training program not only for the faculty and staff of the University but also for others. Betty Love is currently Managing Editor of *Technology.* She has published several papers in peer-reviewed journals; has edited a number of compendia; and is the primary author of *Manual for Public and Stakeholder Participation.* Betty Love received a B.S. degree in Business Administration from Virginia State University in Petersburg, VA, and an M.S. degree in Developmental Clinical Psychology from Antioch College in Yellow Springs, OH.

Jeffrey A. Marqusee* is currently the Technical Director of the Strategic Environmental Research and Development Program (SERDP), and the Director of the Environmental Security Technology Certification Program (ESTCP). SERDP is a tri-agency (U.S. Department of Defense [DOD], U.S. Department of Energy, and U.S. Environmental Protection Agency) environmental research and development program managed by the DOD. SERDP supports research and development to solve environmental issues of relevance to DOD in the areas of cleanup, compliance, conservation and pollution prevention. ESTCP is a DOD-wide program designed to demonstrate innovative environmental technologies at DOD facilities. ESTCP provides for rigorous validation of the cost and performance of new environmental technologies in cooperation with the regulatory and end-user communities. Prior to his current position, Jeffrey Marqusee served as a program manager for environmental technology in the Office of the Deputy Under Secretary of Defense for Environmental Security. He was the principal advisor to the Deputy Under Secretary on environmental technology issues. Before joining DOD, he worked at the Institute for Defense Analyses, where he advised both DOD and National Aeronautics and Space Administration in the areas of remote sensing, environmental matters and military surveillance. Jeffrey Marqusee has worked at Stanford University, the University of California and the National Institute of Standards and Technology. He has a Ph.D. in Physical Chemistry from the Massachusetts Institute of Technology.

Jerome F. McMahon is currently the owner of the consulting company, McMahon Enterprise Group. Previously, he was the Director of Support Services at the D.C. Cook Nuclear Plant, Bridgman, Michigan. There he was responsible for human resources; site protection services including safety, security, and fire protection; and nuclear training and professional development. Additionally, he was process owner for the Indus/Passport installation computer system; Corrective Action Review Board chairman; and Emergency Director. Prior to his current position, Jerome McMahon was the Director of Quality Assurance/Nuclear Training/Emergency Preparedness at PSEG Nuclear, Hancocks Bridge, New Jersey. He was responsible to the Chief Nuclear Officer and President of the Nuclear Business Unit for overseeing the monitoring of plant activities; program reviews; self-assessments to maintain the quality level of the plants, including oversight and administration of the nuclear review board; and the employee concern program. Collateral duties included acting as Process owner for SAP installation; participating on several due diligence merger and acquisition teams; coordinating site work for IBEW contract renewal; and controlling four buildings with a 35 million direct budget consisting of 250 people, 7 direct reports, while reporting directly to the President. His focus for over ten years has been on nuclear training and emergency preparedness. He was Director of Nuclear Training and Radiological Safety at PSEG Nuclear where he directed the training and radiological corporate programs. He was also the Emergency Response Manager (EOF) and Duty Responder. Other past positions include Director of Nuclear Training, Comanche Peak, TU Electric, Glen Rose, Texas; and Supervisor of Technical Training, Wolf Creek Nuclear Operating Corporation, Burlington, Kansas. Jerome McMahon has held the positions of Non-Licensed and Licensed Operator Training Coordinator; Licensed Control Room Operator; and Shift Supervisor and Control Room Operator at various Power facilities across the U.S. Jerome McMahon is a member of the American Society of Training Development and was appointed to the Training Education Executive Committee. He is also a member of the American Nuclear Society. He has served on the Steering Committee for the Symposium on Training of Nuclear Facility Personnel. Other affiliations include the Society for Human Resource Management; American Quality Assurance; the INPO Academy Council; the Industrial Council for the B.S. Nuclear Technology Degree Program at the

University of North Texas; and the Industrial Advisory Council for the A.S./B.S. Nuclear Technology Degree Program at Wharton County Junior College. Jerome McMahon has held the Senior Reactor Operator License at Waterford-3energy; the Reactor Operator License at Fort Calhoun, Omaha Public Power District; the INPO Certified Instructor certification at Wolf Creek Nuclear Operating Corporation; and the B&W Senior Reactor Operator Certification at Midland I and II, Consumers Power Co. Additionally, he served as a U.S. Navy Electronics Technician ETR2 SS as a Qualified Submarine Reactor Operator and held a top secret clearance. He received a B.S. degree in Nuclear Technology from Thomas A. Edison College; an MBA in Management from Amber University; participated in the U.S. Navy Nuclear Power Program; and participated in the University of Michigan Executive and Executive HR programs respectively.

A. Alan Moghissi* is currently President of the Institute for Regulatory Science (RSI), a non-profit organization dedicated to the idea that societal decisions must be based on best available scientific information. The activities of the Institute include research, scientific assessment, and science education at all levels—particularly the education of minorities. Previously, Alan Moghissi was Associate Vice President for Environmental Health and Safety at Temple University in Philadelphia, PA and Assistant Vice President for Environmental Health and Safety at the University of Maryland at Baltimore. In both positions, he established an environmental health and safety program and resolved a number of relevant existing problems in those institutions. As a charter member of the U.S. Environmental Protection Agency (EPA), he served in a number of capacities, including Director of the Bioenvironmental/Radiological Research Division; Principal Science Advisor for Radiation and Hazardous Materials; and Manager of the Health and Environmental Risk Analysis Program. Alan Moghissi has been affiliated with a number of universities. He was a visiting professor at Georgia Tech and the University of Virginia, and was also affiliated with the University of Nevada and the Catholic University of America. Alan Moghissi's research has dealt with diverse subjects ranging from measurement of pollutants to biological effects of environmental agents. A major segment of his research has been on scientific information upon which laws, regulations, and judicial decisions are based—notably risk assessment. He has published nearly

400 papers, including several books. He is the Editor-in-Chief of *Technology: A Journal of Science Serving Legislative, Regulatory, and Judicial Systems*, which traces its roots to the *Journal of the Franklin Institute*—one of America's oldest continuously published journals of science and technology. Alan Moghissi is a member of the editorial board of several other scientific journals and is active in a number of civic, academic, and scientific organizations. He has served on a number of national and international committees and panels. He is a member of a number of professional societies including the American Society of Mechanical Engineers and is past chair of its Environmental Engineering Division. He is also an academic councilor of the Russian Academy of Engineering. Alan Moghissi received his education at the University of Zurich, Switzerland, and Technical University of Karlsruhe in Germany, where he received a doctorate degree in physical chemistry.

Lawrence C. Mohr, Jr.*, is currently Professor of Medicine, Biometry, and Epidemiology; and Director of the Environmental Biosciences Program at the Medical University of South Carolina. His areas of research and special interest include internal medicine and pulmonary disease—specifically diseases of the chest and respiratory system. An area of particular interest to Lawrence Mohr is environmental medicine, including molecular epidemiology and biomarker applications. He has been involved in studies related to environmental lung disease; pathophysiology; prevention and treatment of high altitude illness; high altitude physiology; risk assessment of environmental hazards and clinical epidemiology. Other areas of considerate interest to Lawrence Mohr are assessment of clinical outcomes; health policy analysis; and international health. This latter area includes: global epidemiology; medical relief operations; and health care in Central and Eastern Europe, as well as medical history—the impact of illness on world leaders. Previously, he held academic appointments as a Teaching Fellow in Medicine at the Uniformed Services University of the Health Sciences in Bethesda, MD. He was Associate Clinical Professor of Medicine and Emergency Medicine at George Washington University, Washington, DC. While in these institutions, he was a staff member of the Medical Support Group for the President of the United States. Lawrence Mohr was on the Medical Staff of Walter Reed Army Medical Center—where he completed his Internship and Residency in Internal Medicine—as well as George

Washington University Hospital, both in Washington, DC. He has held Visiting Professorships at various universities. He served as Visiting Chief Resident at Presbyterian Hospital and Visiting Professor at the School of Nursing, both at Columbia University. Additionally, Lawrence Mohr was Visiting Professor of: William Beaumont Army Medical Center, Tulane University, University of Cincinnati, New York University, Brown University, East Carolina University, and the Mayo Clinic. Lawrence Mohr is a Fellow of the American College of Physicians and the American College of Chest Physicians. He is a member of several professional societies including: the American Federation for Medical Research; the Society for Risk Analysis; and the Wilderness Medical Society. Previously, he was on the Scientific Advisory Board for the Consortium in Environmental Risk Evaluation and the Savannah River Health Information System. He has authored or coauthored more than 60 articles, books, or technical publications. He received an A.B. degree in Chemistry as well as an M.D. degree, both from the University of North Carolina, Chapel Hill. Lawrence Mohr, Jr., is certified by the American Board of Internal Medicine.

Goetz K. Oertel's* career in engineering, physics, chemistry, astronomy, and technical program management spans more than 40 years. He consults for industrial, academic, and governmental organizations in North and South America. As President and CEO of the Association of Universities for Research Astronomy, a nonprofit corporation, he engineered the initiation and completion of two 8-m aperture optical telescopes, and oversaw the Space Telescope Science Institute from before launch, through repair of the "Hubble flaw", to its successful operation. He initiated the conceptional phase of the Next Generation Space Telescope that will succeed Hubble as well as the Advanced Solar Telescope, and he oversaw the completion of ambitious ground-based astronomy facilities. He held technical and management positions in the U.S. Department of Energy, including Director of Defense Waste Management; Acting Manager of the Savannah River Operations Office; Deputy Manager of Albuquerque Operations Office; and Deputy Assistant Secretary for Safety, Health, and Quality Assurance. He had primary responsibility for the congressionally-mandated Defense Waste Management Plan, and for managing the related technology development, operations, and projects. He led the initiation of the Defense Waste Processing Facility, and saw it and the Waste Isolation Pilot Plant through technical, managerial, stakeholder, and political challenges. He was National Aeronautics and Space Administration Space Science Chief and Program Manager, and Aerospace Engineer at Langley. He was a Fellow in the White House with the President's Science Advisor and the Office of Management and Budget's Space and Energy branch. He chaired the Westinghouse West Valley Corporation Technical Advisory Group for high-level nuclear waste vitrification and management before, during, and after that project's successful vitrification campaign. He is a member of the American Physical Society, Sigma Xi, and other professional organizations. He is a Fellow of the American Association for the Advancement of Science. He is Chair or member of boards and committees of the National Research Council; George Mason University; the American Society of Mechanical Engineers; International University Exchange; and Westinghouse West Valley Corporation. He is a founding member of the Editorial Board for "Data Science", the new international on-line journal of Codata. He published numerous peer-reviewed papers and was awarded two patents. Trained as electrical engineer and physicist, he received a Vordiplom in Physics and Chemistry from the University of Kiel while on German industrial and governmental scholarships, and a Ph.D. in Physics from University of Maryland at College Park under a Fulbright scholarship.

Francis J. Patti is currently an independent consultant with expertise in the fields of Civil Engineering and Nuclear Engineering. He was Chief Engineer at Burns & Roe. He has been involved in nine major nuclear power projects, working with hot cell facilities; research reactors; power reactors; radioactive waste facilities; and decommissioning projects, which cover the full range of architect-engineer activities and includes both analytical work and systems designs. His consulting assignments have included working with Princeton Plasma Physics Laboratory on the Tokomak Fusion Test Reactor; Duquesne Light Company on Beaver Valley 2; Public Service of New Hampshire on Seabrook; Gulf States Utilities on the River Bend Nuclear Plant; Toledo Edison on the Davis-Besse Nuclear Power Station; Korea Power Engineering Company; North Carolina Electric Membership Corporation on Catawba Nuclear Units 1 and 2; and the Brookhaven National Laboratory, where he was project leader on engineering a major modification to the Brookhaven Medical Research Reactor—to enhance its capability for boron neutron capture therapy to treat certain types of brain cancer. His research reactor experience includes major participation in the design of the Argonne Advanced Research Reactor, the Army Pulsed Reactor Facility, and the NBS Research Reactor. While on these assignments, Francis Patti reviewed, evaluated, or was involved with: providing welding engineering support at construction sites; air modeling, permitting, hazardous waste issues; waste water treatment systems for power plants; emergency plans; procedures; safety analysis; tritium systems upgrade engineering; architect-engineering capabilities; and the status of piping verification work. He provided guidance and support, as well as made recommendations for writing NSSS and turbine specifications, shielding door specifications; and prudency evaluations. Francis Patti was also involved with various studies including: the decommissioning of Brookhaven Graphite Research Reactor; a method of long term decay heat removal; evaluation of containment liner under post LOCA conditions; the conversion of a nearly complete Russian nuclear power plant to coal firing; plutonium disposition; advanced liquid metal reactor; pressurized water reactor; modular high temperature gas reactor; NASA Plum Brook research reactor; Chemical Agent BZ Demilitarization Facility; Hoffman LaRoche/Medi-Physics; Tritium Recovery Facility; and the initial recovery program following the accident at Three Mile Island, Unit 2. During these various studies, he provided technical overviews on: fuel fabrication; process and instrument diagrams; pipe break and safe shutdown analysis; safety evaluations; radiation shielding issues; recycle facilities; filtered containment vent systems; and tritium target fabrication and recovery. Francis Patti was also involved with the preliminary and/or conceptual design of the Central Fuel Cycle Facility, 600 MWe advanced light water reactor power plant for Westinghouse and the U.S. Department of Energy, and stress analysis of piping systems. He was responsible for the conceptual design report and detailed engineering for decommissioning three 100,000 gallon Radwaste Tanks; High Flux Beam Reactor reports for remote control room building and automated vent valve; detailed engineering of Vehicle Radiation Monitor; and a work plan for closure of landfills. Francis Patti has served as Chair of the local sections of the American Nuclear Society and the American Society of Civil Engineers. He received Civil Engineer of the Year and Distinguished Service Awards from the local branch of the American Society of Civil Engineers.

He is the author or coauthor of 15 technical papers. He holds a B.S. degree in Civil Engineering from Drexel University, Philadelphia, PA, and an M.S. degree in Civil Engineering from the Massachusetts Institute of Technology, Cambridge, MA. He also has a diploma in Nuclear Science and Engineering from the International School of Nuclear Science and Engineering as well as an M.S. degree and Professional Degrees in Nuclear Engineering from Columbia University, New York, NY. Francis Patti is a registered Professional Engineer in New York.

Richard Rosenberg is currently a private consultant with experience in the design, development, manufacturing liaison, testing, field installation, and post installation and repair of special-purpose mechanical equipment. Prior to his retirement, he worked for more than 20 years for General Atomics in a number of positions where he fostered the development of Temperature Gas Cooled Reactor (HTGR) technology. Richard Rosenberg started as Senior Staff Engineer of the Fuel Handling Branch where he was technically responsible for the design of remotely operated fuel handling machines. Subsequently, he became Program Engineer of the Rotating Machinery Branch and the Fuel Handling Branch. He oversaw the design, development, testing, field installation, maintenance, and manufacture of the Fort Saint Vrain (FSV) rotating machinery as well as the fuel handling equipment and control rod drives. Richard Rosenberg advanced to Manager of the Fuel Handling Branch; the Mechanical Equipment Branch; and later became Manager of Systems and Components for FSV services. As Manager of these branches, his responsibilities included technical and administrative tasks pertaining to the design, development, and application of the HTGR rotating machinery and service systems; control rod drives; remotely-operated refueling equipment; and reactor servicing equipment. Richard Rosenberg also served as Principal Engineer on the Reactor Development and Technology Staff for the U.S. Atomic Energy Commission. While there, he provided broad overview monitoring of the design, fabrication, and manufacturing efforts for the Fast Flux Test Facility; sodium-cooled reactor-pressure vessel; and guard vessel. He also worked as a Senior Nuclear Engineer for Advanced Technology Laboratories, and as a Senior Engineer for Westinghouse. He was employed with the Oak Ridge National Laboratory where he designed special purpose remote handling equipment and special purpose chemical processing equipment. Richard Rosenberg held various positions as a volunteer with the American Society of

Mechanical Engineers (ASME). He was elected president of ASME in 1988; has served as a member and Chair of the Board of Governors; and as a member of the Board of Trustees. He has been the recipient of the Dedicated Service Award and the Centennial Medallion as well as San Diego Engineer of the Year. He is affiliated with the Civilian Research and Development Foundation as a member of a team reviewing and recommending proposals for joint Russian and American research projects. He holds patents for the High Temperature, High Pressure Sampling System; the Rod Handling System; and the Shielded Cask. Richard Rosenberg has a B.S. Degree in Mechanical Engineering from the University of Tennessee. He is a registered Mechanical Engineer in Pennsylvania; a registered Nuclear Engineer in California; and a Chartered Engineer in the United Kingdom.

Sorin R. Straja is currently Vice President for Science and Technology of the Institute for Regulatory Science. He has over 20 years of expertise in mathematical modeling and software development as applied in chemical engineering and risk assessment. Previously he served as Assistant Professor of Biostatistics with Temple University, Philadelphia; as Director of the Department of Occupational Health and Safety of Temple University, Philadelphia; and as a chemist with University of Maryland at Baltimore. Sorin Straja has extensive experience in the chemical industry where he worked as a senior R&D consultant with the Chemical and Biochemical Energetics Institute, and as a plant manager with Chemicals Enterprise Dudesti and Plastics Processing Bucharest from Romania. He was an Assistant/Adjunct Professor of Chemical Engineering with the Polytechnic Institute Bucharest. Sorin Straja is the author of two books and 44 scientific papers published in internationally recognized and peer-reviewed journals. He was an editor of *Environment International,* and currently is a contributing editor of *Technology.* Sorin Straja received a Certificate of Appreciation for Teaching from Temple University, the "Nicolae Teclu" Prize of the Romanian Academy, and a Certificate of Appreciation from U.S. Department of Agriculture for significant volunteer contributions. He is a Fellow of the Global Association of Risk Professionals, and a member of the American Chemical Society, American Institute of Chemical Engineers, Society for Risk Analysis, and New York Academy of Sciences. Sorin Straja holds a M.S. in Industrial Chemistry and a Ph.D. in Chemical Engineering both from Polytechnic Institute Bucharest.

Glenn W. Suter, II* is currently Science Advisor at the National Center for Environmental Assessment of the U.S. Environmental Protection Agency (U.S. EPA) in Cincinnati, OH. Previous to his current position, he was at Oak Ridge National Laboratory, initially as Research Associate and gradually rising to Science Leader at the Environment Science Division of the Laboratory. His interest has focused on Ecotoxicology in general and Ecological Risk Assessment in particular. He is one of the developers of the most widely-used methodology for Ecological Risk Assessment. This method has been applied to the impact of pollutants on fish, contaminated soils, production of synthetic fuels, and various other ecosystems. Glenn Suter has lectured widely, both nationally and internationally on Ecological Risk Assessment. He is currently a member of the U.S. EPA's Risk Assessment Forum. He has been a member of numerous panels and has consulted with various governmental agencies and private organizations, including the Council of Environmental Quality. He was a member of the Scientific Review Panel for Savannah River Ecology Laboratory; the National Science Foundation Panel on Decision Making and Valuation for Environmental Policy; and the U.S. EPA Science Advisory Board and Conservation Foundation, Ecosystem Valuation Forum. In addition, he was a member of the International Institute of Applied Systems Analysis Task Force on Risk and Policy Analysis and the Council on Environmental Quality. He was a member of the Board of Directors, for the Society for Environmental Toxicology and Chemistry. Glenn Suter is presently on the Editorial Board of *Environmental Health Perspectives* and *Human and Ecological Risk Assessment.* Previously, he was on the Editorial Board of *Handbook of Environmental Risk Assessment and Management* and *Environmental Toxicology and Chemistry.* Glenn Suter is the author of three books and is author and coauthor of over 200 publications. He received a B.S. degree in Biology from Virginia Polytechnic Institute and a Ph.D. in Ecology from the University of California, Davis.

Cheryl A. Trottier* is currently Chief of the Radiation Protection, Environmental Risk, and Waste Management Branch of the Office of Nuclear Regulatory Research at the U.S. Nuclear Regulatory Commission. In that capacity, she is responsible for the management of research programs and the development of technical bases to support rulemaking.

This includes the development of models for assessing the maximum doses likely to be received from lands and materials cleared from regulatory control; evaluating hydrologic model and parameter uncertainty; the development of realistic parameters for assessing sorption processes in geochemical models; and refining evaluations of radionuclide transport mechanism in the environment. In her 25 years of experience in the field of radiation protection, she has been involved in the management of environmental radiation protection monitoring programs and laboratory measurements, and the emergency preparedness coordination for an electric utility. She was also involved in the areas of materials use regulation oversight; development of regulations; and the development of guidance for use of by-product and special nuclear materials. This included finalization of regulations and the development of regulations to certify the gaseous diffusion plants. Cheryl Trottier serves as one of the U.S. representatives to the Nuclear Energy Agency Committee on Radiation Protection and Public Health. She is a member of the American Nuclear Society, and serves as a member of the Committee on Site Clean-Up Restoration Standards. She received her B.A. degree in Biology from Rutgers University.

Charles O. Velzy* is a consultant in the field of waste treatment and disposal. Previously, he held increasingly responsible positions with the environmental consulting engineering firm, Charles R. Velzy Associates, Inc., becoming President in 1976. In 1987, when Velzy Associates merged with Roy F. Weston, Inc., Charles Velzy became Vice President of Weston, a position which he held until retiring in 1992. He has over 35 years of experience as an environmental engineering consultant specializing in: the analysis of waste management problems; design of wastewater treatment and waste disposal systems; and design of new, retrofit of existing, testing, and permitting of waste combustion facilities. He has authored or co-authored over 80 publications—primarily in the field of solid waste management. He has served on the Science Advisory Board of the U.S. Environmental Protection Agency; as President of the American Society of Mechanical Engineers (ASME); Chair of the ASME Peer Review Committee; and as Treasurer of the American Academy of Environmental Engineers (AAEE). He has served on numerous committees of the ASME, the AAEE, the American National Standards Institute, and the American Society for Testing and Materials. He is a registered professional engineer in New York and eleven other states. Charles Velzy received B.S. degrees in Mechanical and Civil Engineering, and an M.S. in Sanitary Engineering from the University of Illinois at Urbana-Champaign.

*** Member of the ASME Peer Review Committee for Energy and the Environment**